SPIDERS OF THE WORLD

A NATURAL HISTORY

EDITED BY

Norman I. Platnick

CONTRIBUTORS

Gustavo Hormiga, Peter Jäger, Rudy Jocqué,
Norman I. Platnick, Martín J. Ramírez, Robert J. Raven

PRINCETON UNIVERSITY PRESS
PRINCETON AND OXFORD

First published in the United States and Canada in 2020 by
Princeton University Press
41 William Street
Princeton, NJ 08540
press.princeton.edu

Library of Congress Control Number: 2020930986
ISBN: 978-0-691-18885-0

This book was conceived, designed, and produced by
Ivy Press
58 West Street, Brighton BN1 2RA, United Kingdom

Publisher **David Breuer**
Editorial Director **Tom Kitch**
Art Director **James Lawrence**
Project Editors **Caroline Earle, Stephanie Evans**
Design **Wayne Blades**
Picture Research **Sharon Dortenzio**
Illustrator **John Woodcock**

Cover photos: Bryce McQuillan: FC left row, fourth down (*Porrhothele*). **Shutterstock**/3DMI: FC center row, top (*Missulena occatoria*); aaltair: FC right row, third down (*Eresus cinnaberinus*); Anton Kozyrev: FC left row, second down (*Steatoda albomaculata*), FC center row, second down (*Misumena vatia*), FC center row, bottom (*Pardosa lugubris*); alslutsky: spine (*Araneus triguttatus*), FC left row, bottom (*Philodromus margaritatus*), FC center row, third down (*Linyphia hortensis*); Dejan Stanisavljevic: FC right row, bottom (*Brachypelma boehmei*); Eric Isselee: FC right row, second down (*Nephila inaurata madagascariensis*); Moisieiev Igor: FC left row, third down (*Argiope bruennichi*); Niney Azman: back cover (*Heteropoda venatoria*); Protasov AN: FC right row, top (Latrodectus tredecimguttatus); Sarah2: FC left row, top (*Zoropsis spinimana*).

Printed in Singapore

10 9 8 7 6 5 4 3 2 1

CONTENTS

INTRODUCTION

BELOW | An orb-web-weaving spider (family Araneidae) in hunting position at the center of its web; orb webs (webs with radii—like a wheel with spokes) are just one of many kinds of webs spun by spiders.

Spiders are among the dominant predators in almost all terrestrial ecosystems, and play a major role in controlling insect populations: the global spider population has been estimated to consume between 400 million and 800 million tons of prey annually. Found on all continents except Antarctica, spiders live in incredibly diverse habitats, from deep in caves to an altitude of almost 22,000 feet (6,700 m) on Mt. Everest (where they survive on prey blown up from lower elevations), and from deserts as hot as Death Valley to tundra in the coldest parts of Siberia. They can even "balloon" into the air, emitting a strand of silk that gets caught by the wind; ballooning spiders have been found alighting on ships more than a thousand miles from the nearest land, and floating at 16,000 feet (4, 900 m) in the air. The many different kinds of silk threads they spin can have a tensile strength greater than steel. Researchers are seeking ways to manufacture textiles that are similarly strong and lightweight for use in products ranging from parachutes to bulletproof vests. Spider venoms are also diverse. They can inhibit the transfer of nerve impulses across synapses, and so they are being studied as possible treatments for diseases such as epilepsy. This book explores the diversity and natural history of these fascinating creatures.

ABOVE | An anterior view of a jumping spider (family Salticidae).

WHAT ARE SPIDERS?

All spiders are arthropods (i.e., members of the group Arthropoda). Arthropods share some striking features, including, most obviously, an exoskeleton. Unlike vertebrates, which have bones inside their bodies, arthropods have an external cuticle that is rigid; in order to grow in size, arthropods must molt periodically (that is, shed their old exoskeletons, and grow new, larger ones). Arthropod bodies are also segmented, with at least some segments bearing jointed appendages. Arthropods are the dominant group of animals on the planet, both in their number of species and in their biomass (i.e., the combined weight of all individuals).

SPIDER RELATIVES

Examples of spider relatives (i.e., other groups of arachnids): **A.** Tailless whip scorpion (order Amblypygi). **B.** Mite (order Acari). **C.** Harvestman (order Opiliones). **D.** Male ricinuleid (order Ricinulei)—note the modification of the third legs. **E.** Pseudoscorpion (order Pseudoscorpiones). **F.** Camel spider (order Solifugae). **G.** Scorpion (order Scorpiones). **H.** Whip scorpion (order Thelyphonida).

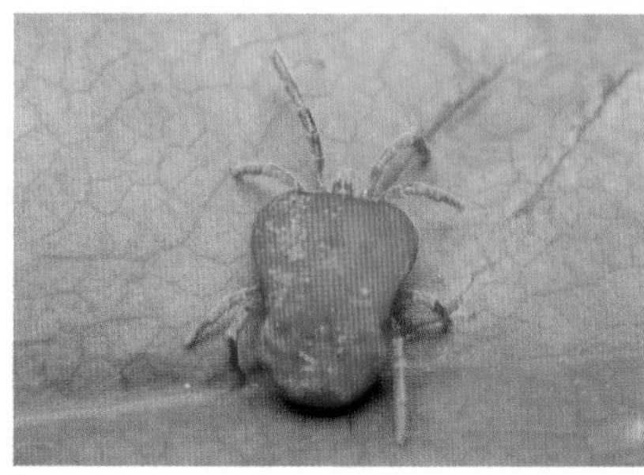

B.

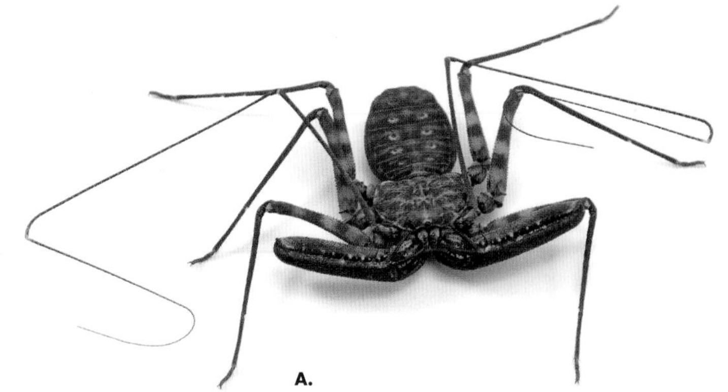

A.

C.

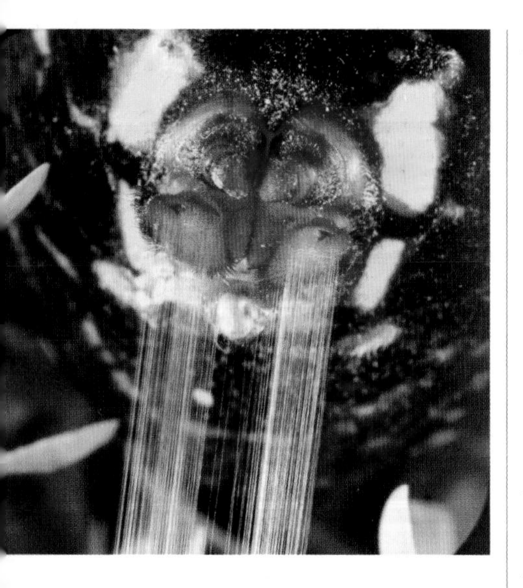

ABOVE | Silk being pulled from the spigots. Abdominal silk glands opening through spigots on the spinnerets are features found only in spiders. The silk is liquid inside the silk glands, but it solidifies as it is being pulled from the spigots.

Among the major subgroups of arthropods, spiders belong not with insects (e.g., beetles or butterflies), crustaceans (e.g., crabs or shrimps), or myriapods (e.g., millipedes or centipedes), but rather with the other chelicerates. Unlike insects, chelicerates have two main body regions (rather than three), four pairs of walking legs (rather than three pairs), and two pairs of mouthpart appendages, the pedipalps and chelicerae (jointed jaws). Living chelicerates include marine animals like horseshoe crabs and (possibly) sea spiders (pycnogonids), as well as arachnids, the group including scorpions, harvestmen, mites, and spiders, as well as some less well-known groups (see photographs above).

All spiders are united as a single group (the order Araneae) because they all share at least two features that are found in no other organisms. The first is abdominal silk glands, which produce silk through spigots carried on the spinnerets, appendages near the back of the abdomen. All spiders produce silk, though only about half the species use that silk to spin webs for prey capture. There are other organisms that produce silk, including

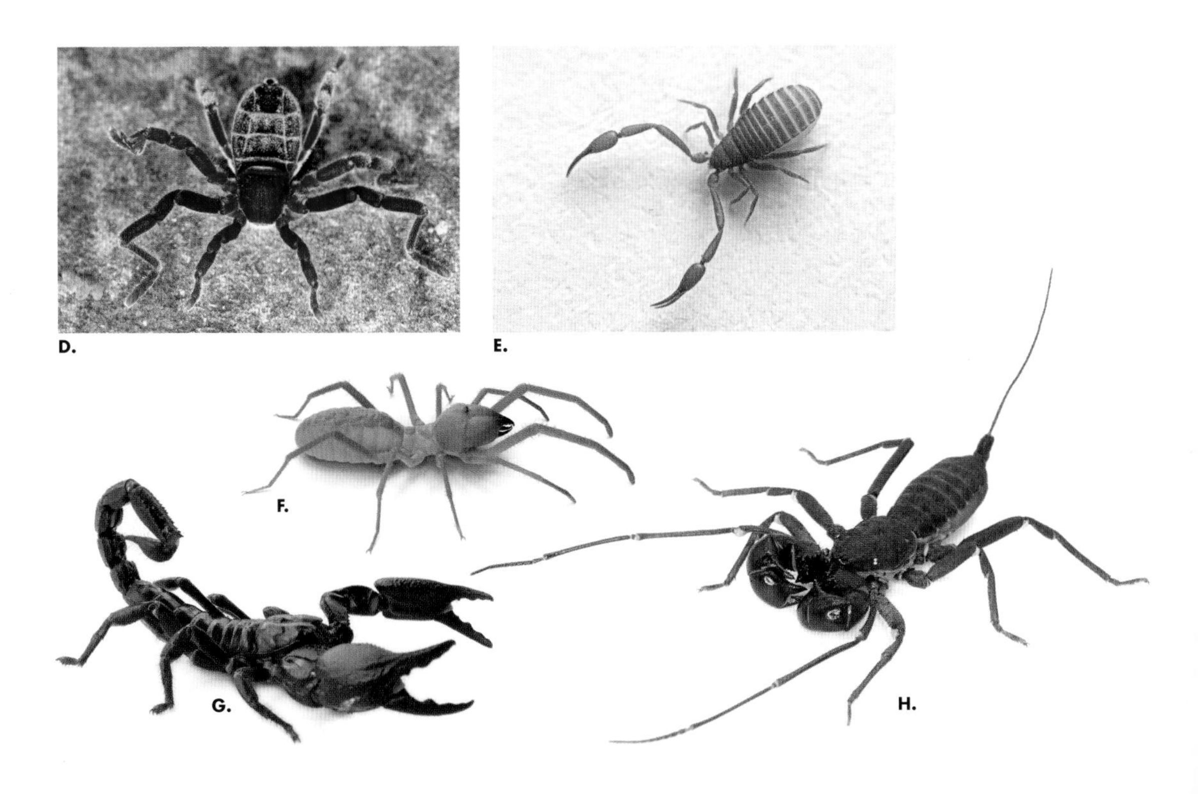

silk moths and some other arachnids, but none of these have abdominal silk glands or spinnerets.

The second unique feature is the modified pedipalps of males. Spiders are also highly unusual, among sexually reproducing animals, in that there is no anatomical connection between the testes, where the sperm are produced, and the intromittent organ used to transfer the sperm to the female. In female spiders, the pedipalps are small and leglike, but in adult males the terminal segment of the pedipalps is modified into a highly complex organ used in mating. A male spider spins a small, dense sperm web, and deposits a drop of sperm from his abdomen onto that silk; the tips of the pedipalps are then dipped into the drop of sperm, and the sperm are drawn into the complex palpal bulb. Later, during mating, the sperm are expelled from the palpal bulb into the reproductive tract of the female. Interestingly, the males of another small group of arachnids, the ricinuleids, use a similarly complex structure to transfer sperm, but this is located on the third pair of legs, not the pedipalps (see photo D above).

ABOVE | A mating pair of forest huntsman spiders (*Heteropoda tetrica*) from Laos; note the translucent sac on the male palp, which expands to help expel sperm from the palp into the female's genitalia.

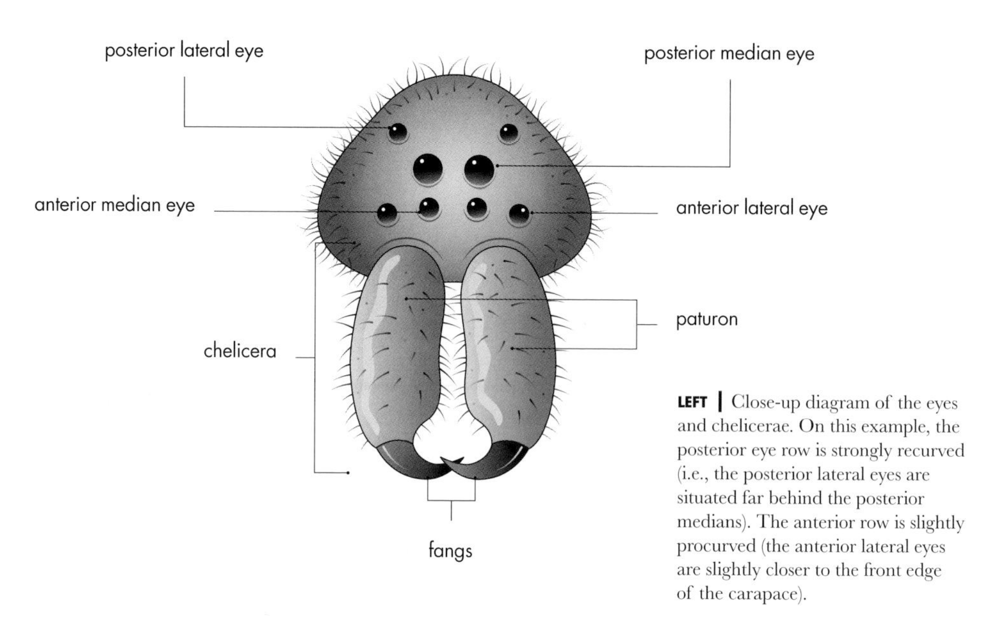

LEFT | Close-up diagram of the eyes and chelicerae. On this example, the posterior eye row is strongly recurved (i.e., the posterior lateral eyes are situated far behind the posterior medians). The anterior row is slightly procurved (the anterior lateral eyes are slightly closer to the front edge of the carapace).

ABOVE | An anterior view of a wolf spider (family Lycosidae), showing the enlarged posterior median eyes.

SPIDER ANATOMY

The body of a spider consists of an anterior (front) cephalothorax and a posterior (back) abdomen, sometimes also called the prosoma and opisthosoma. The cephalothorax and abdomen are connected by a narrow, waistlike pedicel (developmentally, the first abdominal segment). The cephalothorax is covered by two hard plates, a dorsal (top) carapace and a ventral (bottom) sternum. Strong muscles connecting the carapace and sternum can contract quickly, forcing hemolymph (bloodlike fluid) into the legs and thereby extending them. Spider legs have flexor muscles only at some important joints, and the animals have an open circulatory system, without arteries or veins, in which the hemolymph simply surrounds all the internal organs. Near the front, the carapace bears the eyes, usually eight in number and all simple (with a single lens each) rather than compound (with multiple lenses on each eye). To the sides, the cephalothorax bears four pairs of walking legs, each consisting of seven articles (from the body outward, the coxa, trochanter, femur, patella, tibia, metatarsus, and tarsus).

DORSAL VIEW — MALE

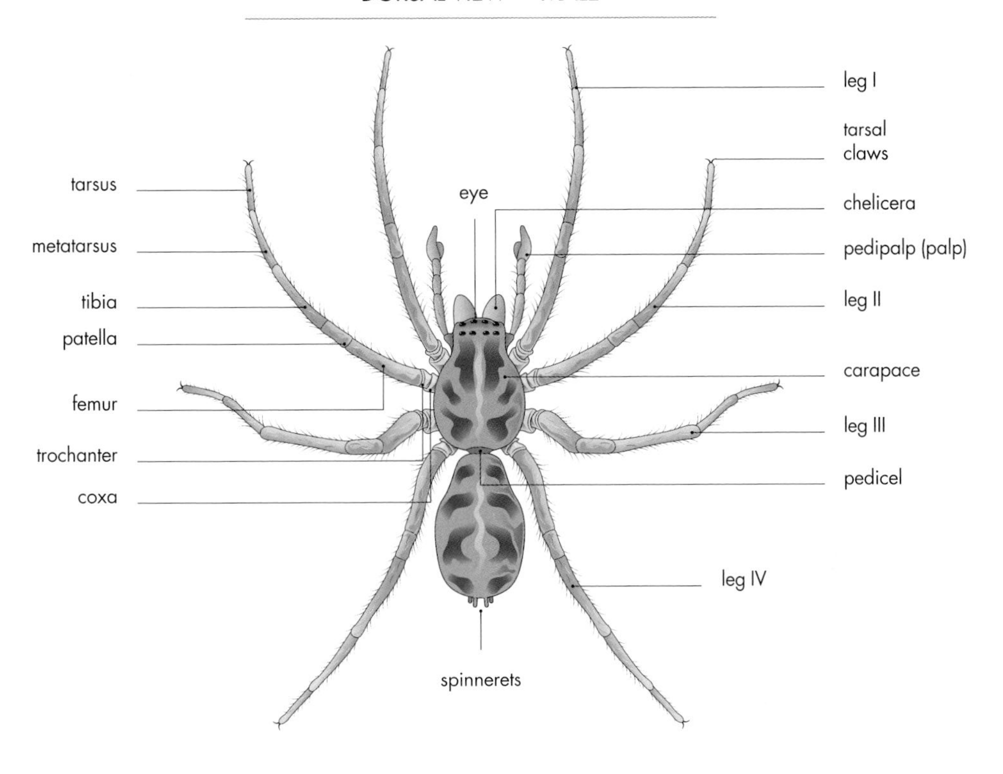

VENTRAL VIEW — FEMALE

RIGHT | Generalized morphology of a spider.

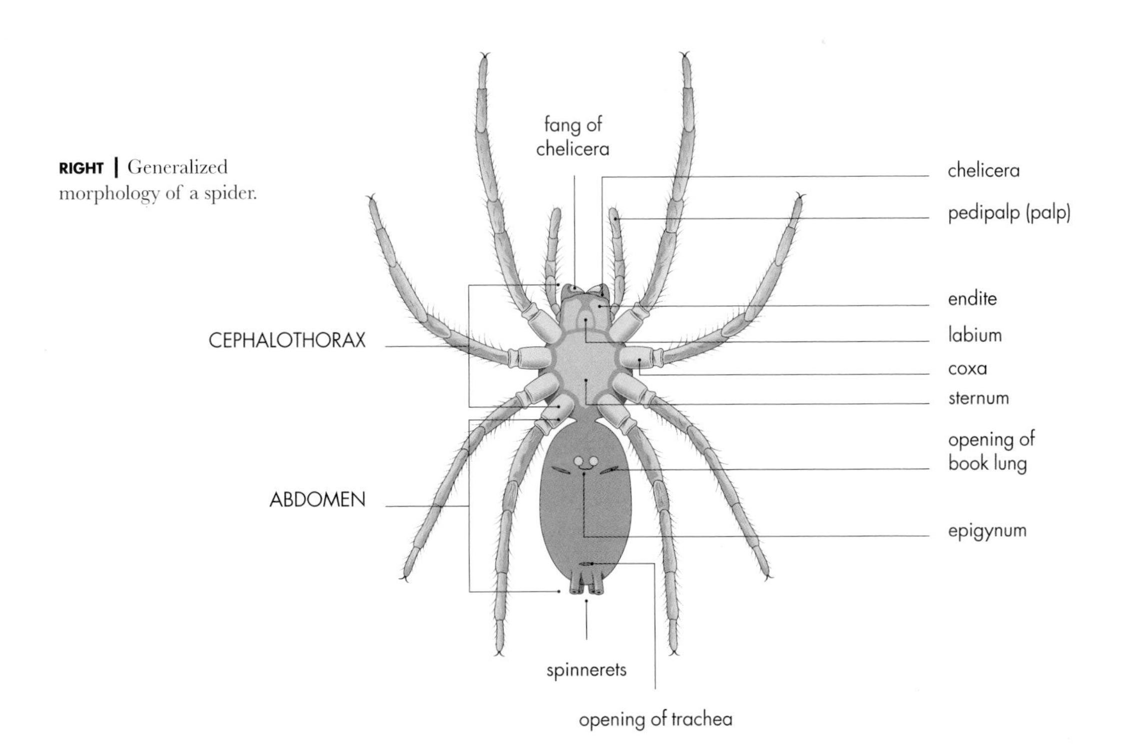

A.

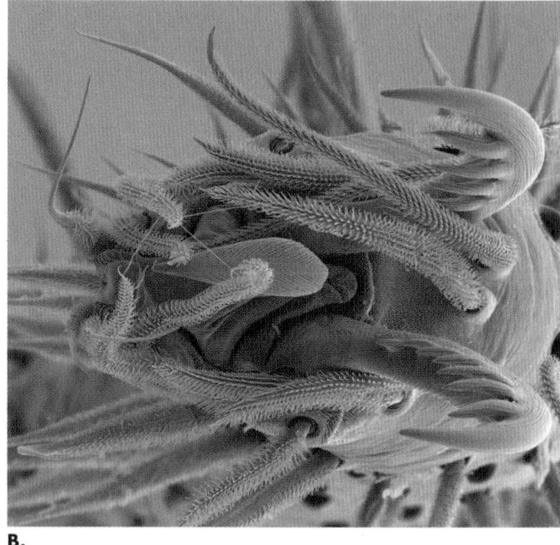

B.

C.

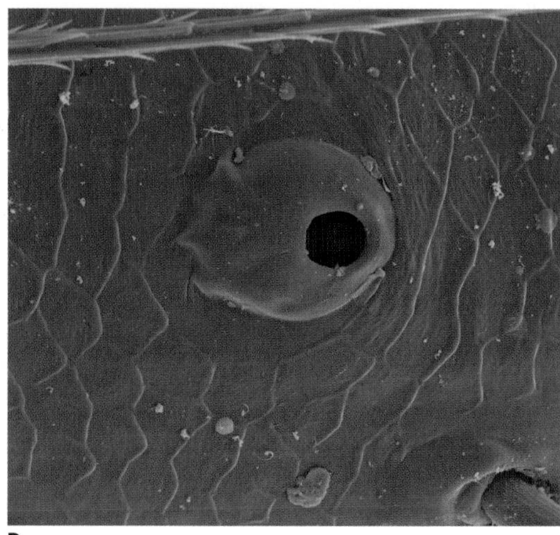

D.

ABOVE | Scanning electron microscope (SEM) images highlighting aspects of spider anatomy: **A**. Leg tarsus with claws, side view. **B.** Tarsal claws, apical view. **C.** Claw tuft setae, side view. **D.** Tarsal organ, dorsal view.

In front of the legs is a pair of leglike pedipalps (palps for short), which have five articles (the trochanter and the metatarsus are lacking). The tips of the last leg segment, the tarsus, have two or three claws (A), usually bearing teeth (B). Two-clawed spiders typically have claw tufts, clumps of specialized setae at their tip; these special setae bear many tiny processes (C) that help the animals adhere even to smooth, vertical surfaces. The leg tarsi usually bear a tiny tarsal organ (D), a chemosensor (a specialized sensory receptor analogous to a nose), as well as trichobothria (E)—long hairs that are extremely sensitive to vibration).

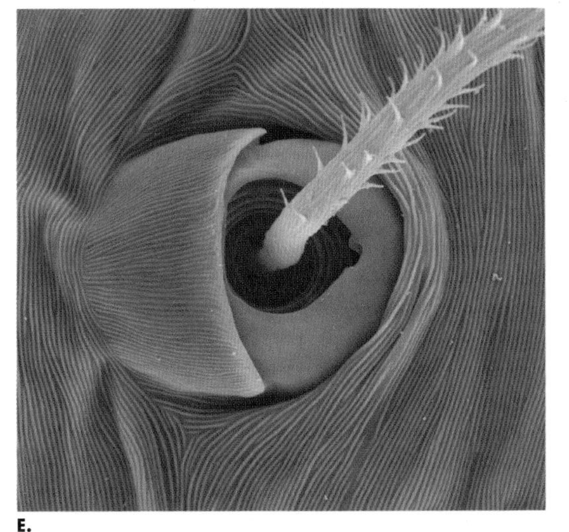

E.

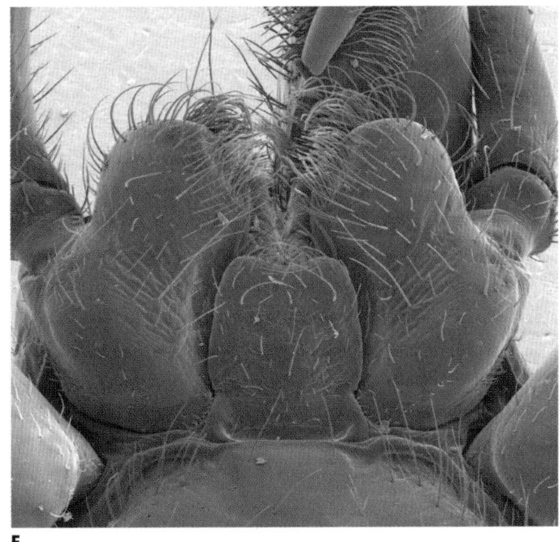

F.

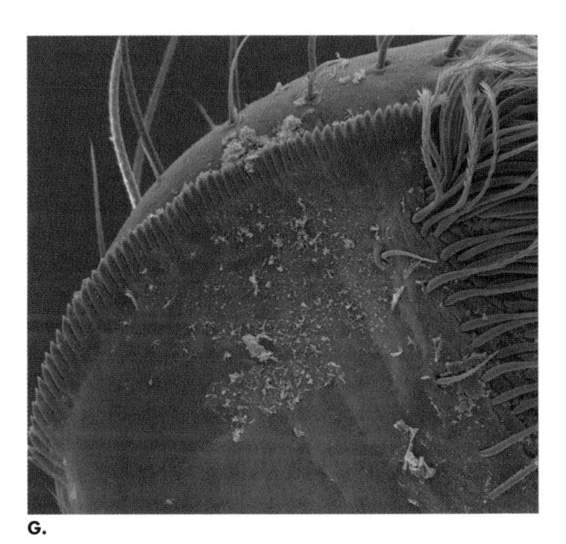

G.

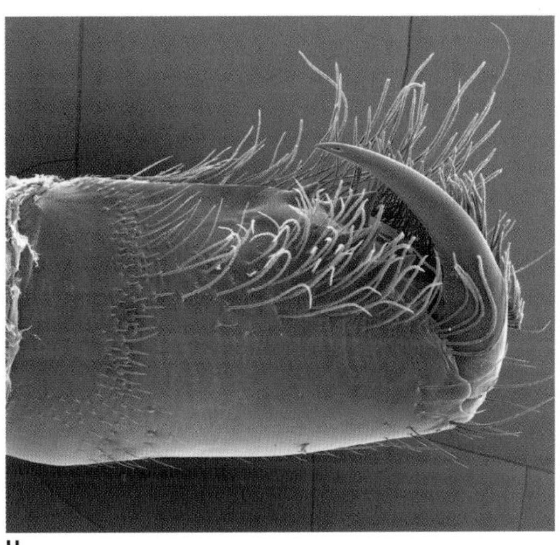

H.

In front of the sternum, a small sclerite (hardened body part), the labium, is flanked by a pair of endites. These are the modified coxae of the palps (F) and usually bear a serrula—a row of teeth on their anterior surface, used for straining food (G). Anterior of the palps are the mouthparts: a labrum, usually hidden by the chelicerae immediately in front of it. The chelicerae bear a ventral fang, and the venom glands open through a small oval opening on the back side of each fang (H). The mouth, through which digestive enzymes are expelled onto prey to dissolve its tissues, and through which the digested fluids are sucked back up, is situated between the bases of the labium and labrum.

ABOVE | **E**. Base of trichobothrium, with basal part of trichobothrial hair, dorsal view. **F.** Labium and endites, ventral view. **G.** Serrula on endite, anterior view. **H.** Chelicera, posterior view.

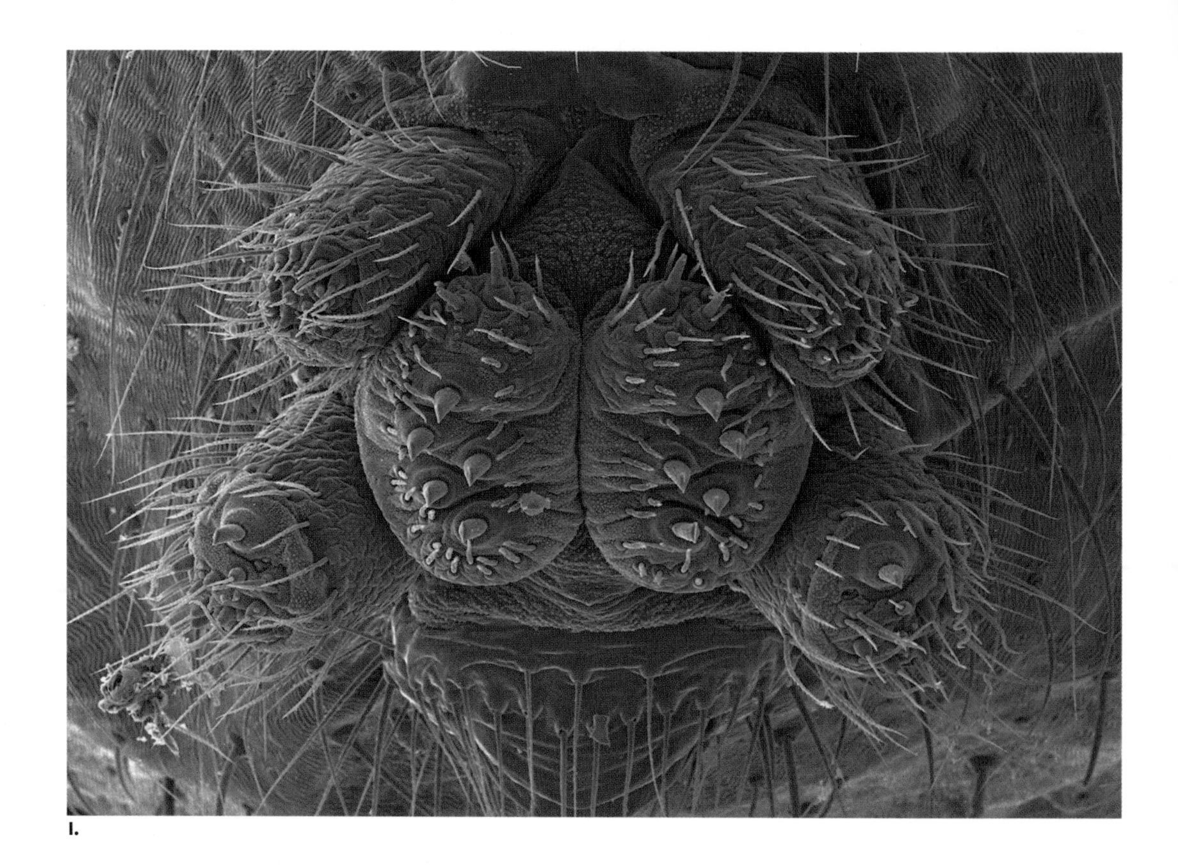

I.

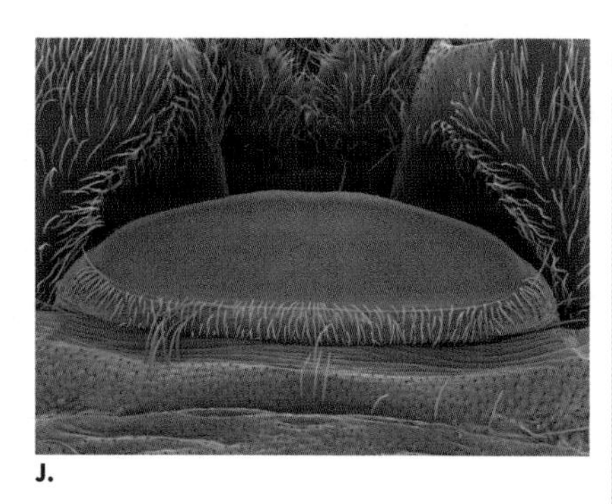

J.

ABOVE | **I.** Spinnerets, ventral view. **J.** Cribellum, ventral view.

The abdomen is generally much softer than the cephalothorax, with most of its features situated on the ventral surface. Anteriorly, there are typically one or two pairs of spiracles that lead to the book lungs, the larger of the animal's respiratory structures. In many cases, spiders also have tubular respiratory tracheae, which most often open through another spiracle just in front of the spinnerets. Most commonly there are six spinnerets, equipped with spigots through which the silk is emitted (I). An individual spider can have as many as seven different kinds of silk glands, each of which open through particular kinds of spigots on particular spinnerets. The most primitive spiders have eight spinnerets, but in most spiders the anterior median pair have fused. When the resulting median structure is functional, it is termed a cribellum, with distinctive silk glands and spigots (J). When the cribellum is functional, it is accompanied by a series of modified hairs on the fourth legs, the calamistrum, used to comb the cribellate silk from those spigots (K). Some adult male cribellates lack spigots on

K.

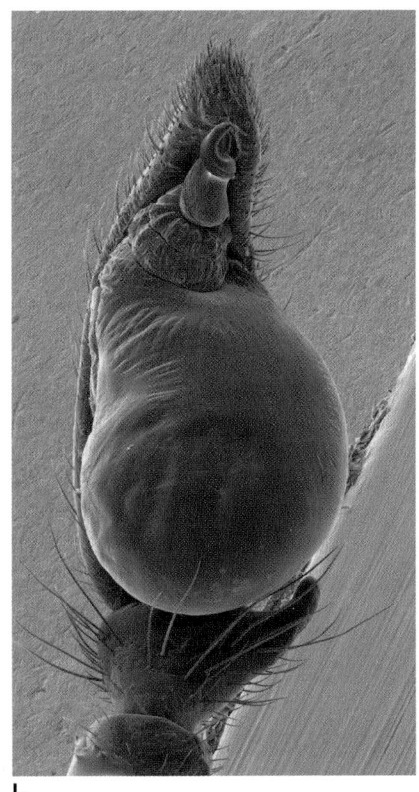

L.

LEFT | **K.** Calamistrum—modified hairs (setae) on the fourth leg, side view. **L.** Male palpal bulb, ventral view.

BELOW | A pair of tree-dwelling trapdoor spiders mating. The male is on the left, and has the tip of one of his pedipalps inserted into the female's reproductive organs.

the cribellum and produce no cribellate silk. In the true spiders that lack a functional cribellum in both sexes, that structure is represented only by a small cuticular lobe with no spigots, the colulus.

In females, the reproductive organs are situated anteriorly in the abdomen. The ovaries produce eggs, which are emitted through the epigastric furrow, an opening extending between the book lungs. In most cases, the females have a complex epigynum (hardened plate that bears the copulatory opening or openings) just in front of the epigastric furrow; during mating, the tips of the male palpal bulbs (L) are inserted into the epigynum, and the sperm are transferred into the internal storage receptacles (spermathecae) of the female. The intricacies of the male palp and female epigynum often provide the best evidence for distinguishing closely related species (see pages 165–167).

THE SPIDER FAMILIES

SPIDER CLASSIFICATION

The diversity of organisms on our planet is astounding. Since long before Charles Darwin, biologists have tried to organize that diversity by classifying individual organisms into species and groups of species, a field called taxonomy. When biologists assert that there is a group, such as spiders, they are asserting that every species placed in the group is more closely related to the other members of the group than to any organisms placed outside the group. If we were to find convincing evidence that some particular species currently considered a spider is actually related more closely to, say, a scorpion or a harvestman, then our hypothesis about spiders as a group would require revision. The same is true for all the subgroups of spiders that we recognize, and all the larger groups to which spiders belong.

ORDER
ARANEAE
SPIDERS

SUBORDER
MESOTHELAE
Spiders with a series of plates on the abdomen, and with the spinnerets situated near the front of the abdomen—contains just one spider family: Liphistiidae.

SUBORDER
OPISTHOTHELAE
Spiders without a series of plates on their abdomen, and with the spinnerets situated near the rear of the abdomen—contains all the other families.

Spiders currently comprise more than 48,000 species, placed in more than 4,000 genera and 115 families. This book explores and celebrates that diversity. It provides photographs and natural history information for representative genera and species, arranged by family and accompanied by maps of their geographic distribution, diagrams, and tables of their salient characteristics.

The family-level classification of spiders is an area of active research. For our purposes, during the production of this volume we chose to adopt the most widely accepted current list of families, as presented in the World Spider Catalog (WSC) at the onset of our work. During production, some newly proposed changes in the list have been adopted by the compilers of the WSC. We have made no attempt to keep up with those changes; many are ephemeral and will disappear as soon as the next papers on those particular spiders appear in the published literature. Even if we had attempted, and succeeded, in having our family list match that of the WSC on the book's day of publication, that matching would disappear within a few months, if not weeks.

Relationships among the spider families are also an area of active research and substantial disagreement. For our purposes, we have therefore recognized only some major groupings of families that are widely accepted by almost all researchers in the field. So, for example, spiders (the order Araneae) are generally divided into two suborders, called the Mesothelae and Opisthothelae. The Mesothelae includes only a single family, the Liphistiidae; all the other families are placed in the Opisthothelae. Our first family treatment thus details the Liphistiidae (pages 16–17), and is followed by a section detailing the features that separate all the families of the Opisthothelae from liphistiids. Similarly, the Opisthothelae is generally divided into two smaller groups (infraorders), the Mygalomorphae (containing tarantulas and their relatives) and the Araneomorphae (the more typical spiders). Thus, our discussion of the Opisthothelae also focuses on the features

INFRAORDER
MYGALOMORPHAE
Tarantulas and their relatives; 20 families—chelicerae move with an up-and-down motion.

INFRAORDER
ARANEOMORPHAE
"True" (typical) spiders; contains the majority of the spider families, genera, and species—chelicerae move from side to side.

Araneomorphae consists of several subgroups, including:

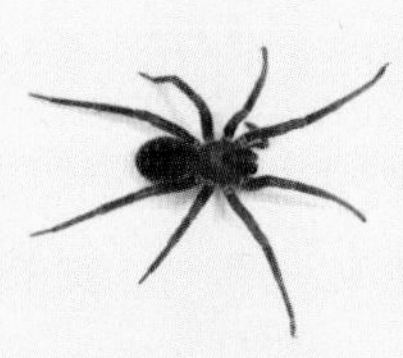

SUBGROUP
AUSTROCHILOIDEA
Small group of two families in the austral continents; most have four book lungs, as in tarantulas.

SUBGROUP
PALPIMANOIDEA
Assassin spiders and their relatives; five families with peg teeth on their chelicerae.

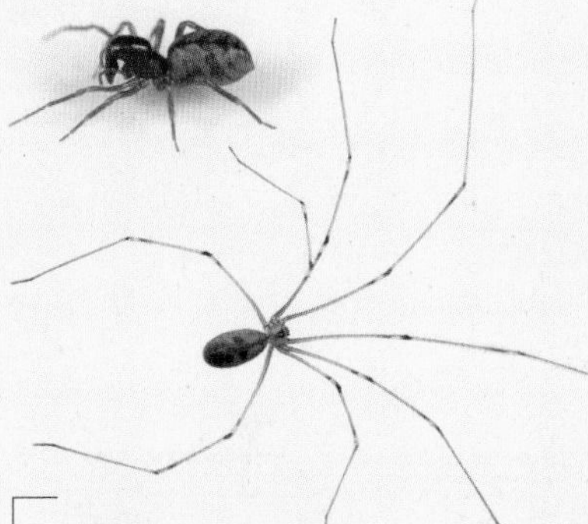

SUBGROUP
SYNSPERMIATA
Brown recluses, cellar spiders, and their relatives; a large group of spiders producing fused sperm cells and having simple male genitalia.

SUBGROUP
ENTELEGYNAE
From the Greek, meaning "females with complete genitalia." A large group of families—more than 70—that share a special arrangement of female genitalia.

CLADE
"RTA CLADE"
40 families of entelegyne araneomorphs. The only character that unites them is the presence of a retrolateral tibial apophysis (RTA), a projection on the tibia of the male palp.

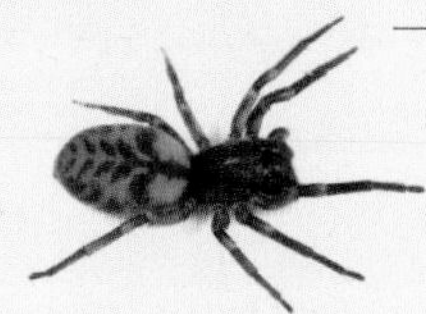

that unite the various mygalomorph families. Subsequent discussions of widely accepted groups detail the features uniting the araneomorphs, the differences between haplogyne and entelegyne araneomorphs, and the clustering of some entelegyne families into larger groups (such as the "RTA-clade").

Selecting the genera to use as examples for each of the families we treat has been challenging. For most families there is what we could call typical behavior: Araneidae live in orb webs, Lycosidae are diurnal hunters, Pisauridae construct tridimensional webs. We therefore often chose example genera that show such typical behaviors. But in almost all families there are also exceptions: some Araneidae make a completely different kind of web, there are night-active Lycosidae (some of which live in a funnel web), and the most widespread pisaurid is a hunter that only constructs a web for its offspring. It was impossible to show all the exceptions, but some of them are so striking that we could not refrain from detailing their habits, even though they deviate from what can usually be expected for that particular family. In the family discussions, the maps provided show the distribution of the genus in the heading, not the entire family.

LIPHISTIIDAE: *LIPHISTIUS*

SEGMENTED TRAPDOOR SPIDERS

LEFT | A female *Liphistius erawan*, a species known only from around the Erawan Waterfall in the Maeklong Mountains of western Thailand.

RIGHT | *Liphistius desultor*, a large species found on Penang Island, Malaysia, shares its strikingly bicolored legs with a species from Sumatra.

Liphistiids are a group of eight genera, including around a hundred species, that together represent the sister group of all other living spiders. They constitute the suborder Mesothelae, whereas all other known extant spiders (more than 48,000 currently valid species) belong to the suborder Opisthothelae. Sometimes referred to as "living fossils," liphistiids retain a number of features that seem to be primitive for spiders, including a series of hard transverse plates on the top of the abdomen that reflect the original segmentation of the body. Members of *Liphistius* retain the original complement of eight separate spinnerets, and these are positioned much closer to the front of the abdomen than in other spiders. Like other trapdoor spiders, they spend almost their entire lives inside a tubular

GENUS
Liphistius

DISTRIBUTION
Myanmar, Laos, Thailand, Malaysia, and Sumatra

HABITAT
Tubular burrow lined with silk, capped with a silk trapdoor

CHARACTERISTICS
- Series of plates on top of abdomen
- Constructs a well-camouflaged trapdoor
- Spinnerets are near front of abdomen

silk-lined burrow that is capped with a trapdoor made of silk, camouflaged with bits of debris from the surrounding leaf litter; when closed, the trapdoor can be very difficult to see.

In *Liphistius*, a series of six to eight silken "fishing lines" radiate from the rim of the burrow. When the spider is hungry, it rests at the top of the burrow, with the trapdoor slightly ajar and the tip of a leg positioned on each of the triplines. When the animal senses vibrations from any passing insect that disturbs one of the triplines, it quickly dashes out, captures the unfortunate insect, and carries it back inside the burrow to consume. In the other liphistiid genera, found in China, Vietnam, and Japan, there are only seven spinnerets, and the burrows lack triplines.

OPISTHOTHELAE AND MYGALOMORPHAE

Within the suborder Opisthothelae, two smaller groups (infraorders) are recognized, the Mygalomorphae (tarantula-like spiders) and the Araneomorphae (see pages 54–57), usually known as the "true spiders" ("true" here means "typical" spiders, and not that liphistiids and mygalomorphs are in any sense "false" spiders). The many spiders grouped in the Opisthothelae are united by several features that generally represent modifications of the more primitive character states found in the Liphistiidae (discussed on the previous two pages)—for example, placement of the spinnerets near the end of the abdomen, and the loss of a separate ventral abdominal sclerite carrying the posterior pair of book lungs.

LEFT | A Greenbottle Blue Tarantula (*Chromatopelma cyaneopubescens*). The New World and Asian tarantulas upstage those from Europe, Africa, and Australia in terms of their size and color. In addition, 90 percent of the New World species have highly irritating defensive hairs on their abdomen.

CHELICERAE MOVEMENT DIFFERENCES

In mygalomorphs and liphistiids the fangs extend longitudinally and the chelicerae move in an up-and-down motion. In araneomorphs, the fangs usually orient toward the midline and the chelicerae move from side to side.

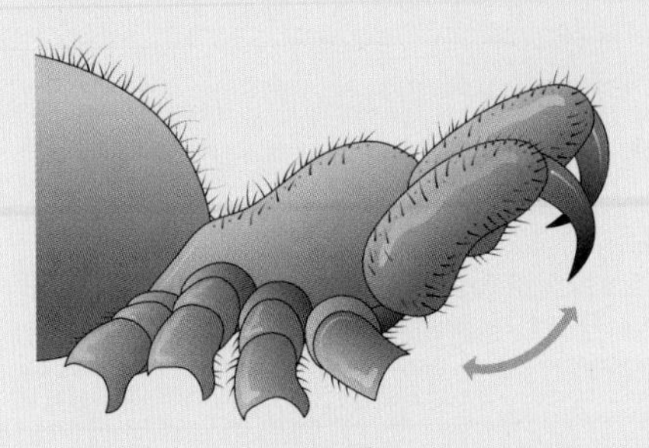

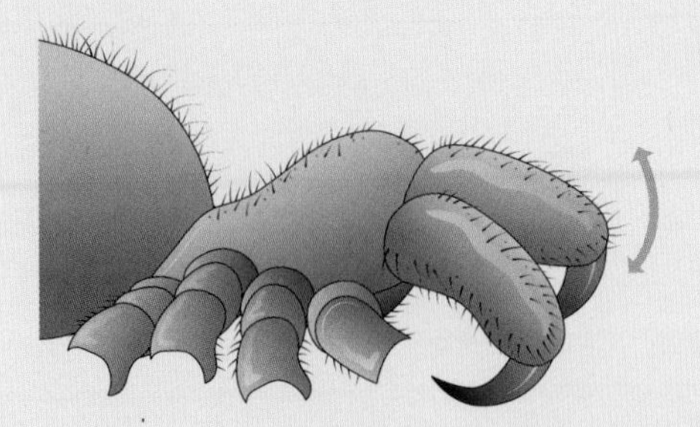

LIPHISTIIDS AND MYGALOMORPHS

Chelicerae move up and down.

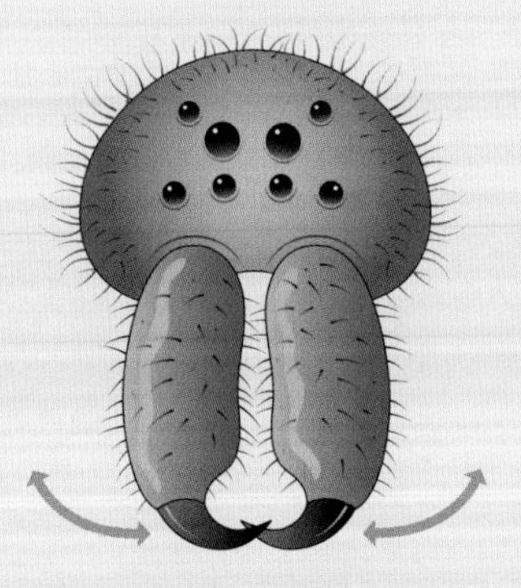

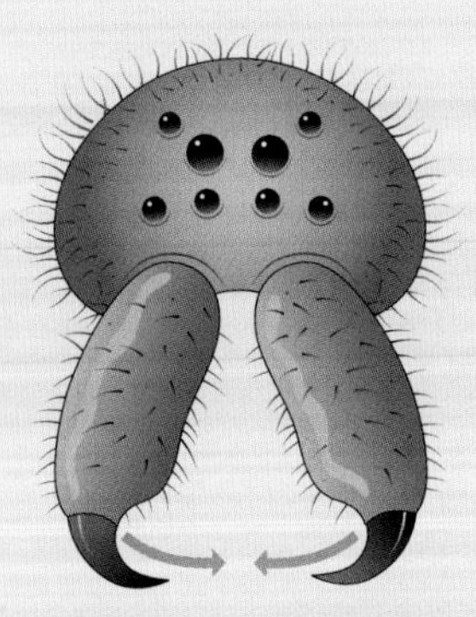

ARANEOMORPHS

Chelicerae move from side to side.

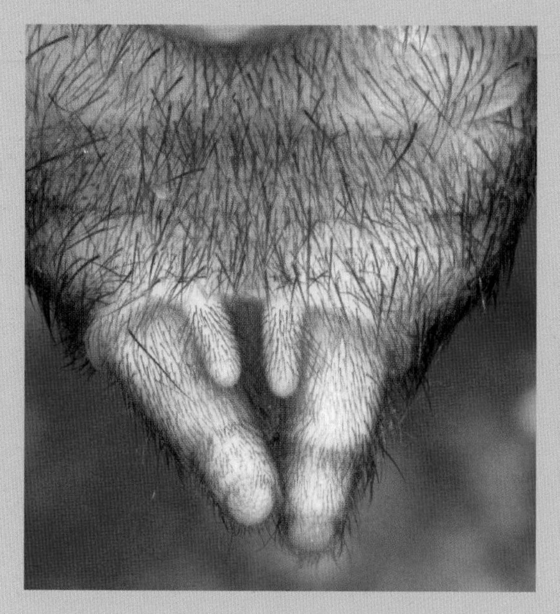

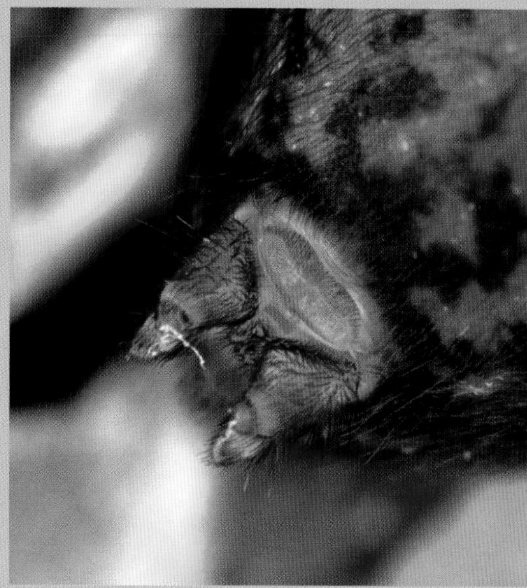

ABOVE | Left, the four spinnerets of an idiopid mygalomorph: as in many mygalomorphs the anterior lateral spinnerets are absent; the inner smaller pair (posterior median) are barely functional and lost in some genera; the larger (posterior lateral) spinnerets have the domed terminal joint characteristic of a group of families. Right, the spinnerets of an araneomorph; the anterior median spinnerets of this species have fused to produce a cribellum, the broad plate in front of the other six spinnerets.

Some authors have argued that liphistiids are most closely related either to all, or to just some, mygalomorphs, as there are several characters in which mygalomorphs resemble liphistiids. For example, both groups have chelicerae that move mostly in an up-and-down plane. In contrast, araneomorphs have chelicerae that move mostly with a side-to-side motion. Is this evidence that mygalomorphs are more closely related to liphistiids than they are to araneomorphs?

For such characters to qualify as evidence, the similarities under discussion have to be unique to the group in question (implying that they first evolved in the most recent common ancestor of the group and have subsequently been inherited by all of that ancestor's descendants). In this case, chelicerae with an up-and-down motion do not qualify: they are not unique to a group containing just liphistiids and mygalomorphs. Chelicerae with this kind of motion are also found in the closest relatives of spiders (e.g., other arachnid groups such as the tailless whip scorpions), implying that they first evolved in some earlier common ancestor of spiders and those other arachnid groups. In other words, for spiders, chelicerae with an up-and-down motion are the "primitive" (i.e., original) kind of chelicerae, and the character actually provides evidence for grouping together all the araneomorphs, which share a modification of that primitive condition.

Mygalomorphs include the next 20 families discussed here; they are unique in having lost the anterior median pair of spinnerets (even in their early developmental stages, prior to hatching). In araneomorphs, that pair of spinnerets does appear embryologically, as it does in liphistiids, but it fuses during development.

ATYPIDAE: *SPHODROS*

PURSE-WEB SPIDERS

Atypids include three genera, together containing more than 50 species. They are among the most basal groups of mygalomorphs (tarantulas and their allies). Like other mygalomorphs, their chelicerae move with an up-and-down motion (in contrast to araneomorphs, the typical spiders, whose chelicerae move from side to side). Their common name refers to their unique lifestyle. These spiders spend virtually their entire life span (which in females can be more than seven years) within a silken tube extending deep into the soil; the long, narrow tube resembles a type of lady's purse common in the nineteenth century, hence their name "purse-web spiders." The tube also has a portion above ground, which in some species of *Sphodros* extends as much as 20 inches (50 cm) up the side of a tree trunk.

When the spider senses prey walking on the surface of the tube, it rapidly pierces the tube with its long fangs, impaling the prey and injecting paralytic

GENUS
Sphodros

DISTRIBUTION
USA and Canada, from Nebraska, southern Ontario, and Rhode Island south to eastern Texas and northern Florida

HABITAT
Silken tube buried into soil, extending above ground near tree trunk

CHARACTERISTICS
- One plate on top of abdomen
- Inner part of palpal coxa prolonged into a distinct lobe
- Six spinnerets, posterior median pair widened and with triangular tips

ABOVE | Anterior view of a female *Calommata signata* from Japan, showing the elongated chelicerae; note the transverse ledge at the base of the fangs, which is unique to this genus.

LEFT | A male *Sphodros rufipes* from the USA, showing the striking color contrast between the legs and body; in females, the legs do not show this contrasting coloration.

venom. It then cuts a slit in the tube large enough to drag the prey inside, repairs the tear with new silk, and feeds on the hapless insect. Adult males leave their tubes to search for females; when they encounter the aerial portion of a female's tube, they drum on its surface with their legs and palps, communicating their intentions so as not to be taken for prey. The males also exude a large drop of saliva, enabling their digestive enzymes to soften the silk of the female's tube, which they then enter. Mating, like feeding, occurs inside the tube, and in some species the male can continue to live peacefully with the female for several months.

In some species of *Sphodros*, and the similar genus *Atypus*, the aerial portion of the tube is shorter, and extends horizontally along the surface of the ground. The tubes are so well camouflaged that they are seldom seen. Often the presence of atypids in an area is first detected only by catching the wandering adult males that stumble into pitfall traps (open containers sunk into the soil to capture animals moving on the ground).

Atypus species occur widely in Eurasia; there is also one species known only from Pennsylvania, USA. The third genus, *Calommata*, occurs from Africa to Japan. It includes bizarre species with dorsally expanded chelicerae and long fangs bearing a basal ledge on their outer surface.

ANTRODIAETIDAE: *ANTRODIAETUS*

FOLDING TRAPDOOR SPIDERS

BELOW | Female *Antrodiaetus unicolor*—the high head and more slender front legs, coupled with the dark halfmoon plate on its abdomen, are distinctive.

The genus *Antrodiaetus* includes 21 species, two of which occur in Japan and the remainder in North America. They are one of an elite group of mygalomorphs that are burrowers and whose range was under ice during last glaciations in the northern hemisphere. Did they survive under the snow or recolonize areas as the ice retreated? Indeed, among the most ancestral members of the trapdoor groups (Atypoidea), these spiders retain not only the third pair of spinnerets but also one to four remnants of

GENUS
Antrodiaetus

DISTRIBUTION
USA, western Canada, Japan

HABITAT
Tubular burrows, in banks and slopes, or under logs; covered entrances

CHARACTERISTICS
- Short spinnerets, barely evident from above
- Spiny digging process (rastellum) on the chelicerae
- Eight eyes in a trapezoidal group
- Front two pairs of legs weaker than back pairs
- Males have a secondary palpal bulb; females have four separate sperm receptacles internally

the hard plates on the dorsum of the abdomen, representing vestiges of the separate segments that we know so well in scorpions and other arachnid orders. In addition, *Antrodiaetus* have very special male and female genitalia. Some species build a deep burrow in banks and similar slopes, closed with a cryptic (well-camouflaged), floppy entrance—not unlike that of some of the Australian mouse spiders (*Missulena*). Others make filmy tubes under logs, with just a short portion underground.

BELOW | The long palps and vertically acting fangs of *Antrodiaetus* species are diagnostic of the infraorder Mygalomorphae.

MECICOBOTHRIIDAE: *MECICOBOTHRIUM*

MIDGET FUNNEL-WEB BUILDERS

In one species of a strange family that includes a total of four genera and nine species, the diagnostic whiplike posterior lateral pair of spinnerets is strongly reminiscent of the spinnerets in the recently discovered *Chimerarachne*—a fossil spider with a tail that was found in Burmese amber. The associations of the Mecicobothriidae are traditionally with the purse-web spiders (Atypidae) and folding trapdoor spiders (Antrodiaetidae), which have also retained some vestiges of abdominal plates as well as having specially modified sexual organs. However, the mecicobothriids' resemblance to their distant advanced araneomorph relatives has gone two steps further. They have both a shallow, long, cephalic groove (fovea), to which the sucking stomach muscle is attached, and male palpal organs that, taken alone, could easily be mistaken as araneomorph characters.

Unlike the Atypidae, the other families of this special group have conservatively shaped mouthparts, but *Mecicobothrium* shows the beginning

RIGHT | A female of the Californian species *Megahexura fulva*, showing the many-jointed spinnerets (used to make a delicate sheet web) and the halfmoon plate near the front of the abdomen.

LEFT | *Mecicobothrium thorelli*, a tiny litter spider with long spinnerets showing the small halfmoon plate on its abdomen.

GENUS
Mecicobothrium

DISTRIBUTION
Southern Brazil, Uruguay, and northern Argentina

HABITAT
Leaf litter in forests

CHARACTERISTICS

- Three tarsal claws
- Cephalic groove (fovea) longitudinal
- Plates on abdomen
- Male palpal bulb lies in a long modified final joint
- Posterior lateral spinnerets very long, with last joint whiplike

of the development that becomes extreme in the Atypidae. The USA boasts three of the four known genera, with *Mecicobothrium* itself known only from two species occurring in Brazil, Uruguay, and Argentina. Most male mygalomorphs have some mechanism for safely managing the raised and confronting chelicerae and fangs of the female during mating. These leg spurs, lobes, and spines are absent in *Mecicobothrium*, but the bizarrely modified chelicerae of male *Mecicobothrium* serve the same function.

The only member of its genus, *Megahexura fulva* is a tiny yet widespread Californian spider that has largely evaded biological study. The spider uses its delicate, many-jointed long spinnerets to create tiny sheet webs that hang from embankments, and its mouthparts show even more development that those of *Mecicobothrium*. Equally, the tarsus of the legs of males is curved and many jointed, but the chelicerae have no sign of fang-immobilizing mechanisms; perhaps like some Australian tarantulas, the mating spiders lay their legs along the top of each other's head.

ACTINOPODIDAE: *MISSULENA*

MOUSE SPIDERS

One of three quite strange mygalomorph genera that occur only in Chile and Australia, *Missulena* presents a puzzling combination of characters and associated behavior. The genus includes 17 species that are widespread in Australia, as well as one—probably misplaced—species in Chile. Females build highly cryptic burrows of wafer-thin silk that reach down about 1 foot (0.3 m) into the soil. Their most conspicuous behavior involves the very large and powerful muscles in the head that operate the chelicerae and fangs. When disturbed, the spider will rise quickly into a defensive position, with the front legs raised, the chelicerae poised widely apart, and the fangs outspread. As this display presumably suggests to potential enemies, a *Missulena* bite can be ferocious: the spiders often take a full lock with the fangs in the offender/prey and do not release (known as a "pit bull bite").

GENUS
Missulena

DISTRIBUTION
Australia (except for Tasmania), Chile

HABITAT
Gardens plus grasslands, and from deserts to rainforests

CHARACTERISTICS
- Large, shiny black spiders, quick to assume a defensive position
- Four very short spinnerets
- Long conical labium, flanked by rotated endites
- Eyes spread widely across the head
- Three tarsal claws

LEFT | This dashingly colorful *Missulena occatoria* male is active by day and walks under the much larger female raised on tiptoes in her burrow to mate.

ABOVE | The diagonally operating fangs of this female *Missulena bradleyi* grip vise-like and lock on their prey.

Unusually, *Missulena* spiders rarely use the venom from their reduced venom glands but, when delivered, this has a similar toxicity to that of the Australian funnel-webs (family Atracidae), even though it is biochemically dissimilar. Separate bites to an 18-month-old infant and a young father resulted in serious systemic symptoms and unilateral weakness, respectively. Although there is not a specific antivenom for *Missulena*, the antivenom for funnel-web bites is effective, as the two venoms act similarly despite their differences. In contrast, a "dry" or venom-free bite caused only minor pain and puzzlement in a seven-year-old boy before the spider was summarily crushed and the embedded fangs removed.

The cryptic burrows of these spiders often go unnoticed in many Australian lawns until, with the onset of fall, dozens of male *Missulena occatoria* emerge and wander in search of females. The males are brightly colored, with red chelicerae and head contrasting with the black remainder of the cephalothorax and the often royal blue body. Such brightly colored males can be seen queuing at the entrance to a female's burrow in daylight.

MIGIDAE: *HETEROMIGAS*

RIDGE-FANGED TRAPDOOR SPIDERS

Migids are found in South Africa, Madagascar, Australia, New Caledonia, New Zealand, Chile, and Argentina. Given this typical Gondwanan distribution, their apparent absence from the Indian subcontinent remains puzzling, although it is possible that they have been overlooked there—the very small trapdoors might be dismissed as belonging to juvenile idiopids, as sometimes happens in Australia. One of two species of *Heteromigas* makes a short burrow with a thick plug door in flat ground and embankments. Most migids are smaller than *Heteromigas*, which often occurs together with species of *Migas*, but one Malagasy genus (*Thyropoeus*) is a veritable giant, approaching the size of a small tarantula.

Like the ornate barychelid *Sason*, the small, cryptic burrows of these spiders are challenging to find. Their presence on a tree, for example, is

LEFT | A female *Heteromigas* species out of her burrow on moss; these tiny trapdoor spiders possess a high head, weaker front legs, and very short spinnerets behind.

RIGHT | The very short, uncovered silken burrow of *Poecilomigas*, excavated into the soft bark of a tree, makes these species very vulnerable to spider-hunting wasps, whose passage is indicated only by a puparium instead of a spider in the burrow.

GENUS
Heteromigas

DISTRIBUTION
Queensland and Tasmania, Australia

HABITAT
Cooler forests; the burrow is built with a door flush with the ground

CHARACTERISTICS
- No digging process (rastellum) on chelicerae
- Eyes in a widely rectangular group
- Fangs with two keel-like ridges along outer surface

best indicated by the small white doors of empty burrows—hanging by their lower hinge, rather than the upper hinge, as in most trapdoors. These short burrows are built just under thin bark; judging by the prevalence of insect puparia in empty burrows, the inhabitants are easily preyed on by spider wasps. Wasps hunting by tapping across bark rich in spiders will detect sounds indicating a hollow. The wasps then sting the spiders and lay their eggs directly into them.

Migas species in the cooler parts of Tasmania (and one in New Zealand) have taken to building near the sea. These short, flimsy burrows in the bark of trees are thought by some to have provided enough protection for tiny, mountaintop migids in South Africa to get down to the Indian Ocean and migrate, presumably via rafts of wood or tree bases, to Kangaroo Island in South Australia.

Poecilomigas builds a fragile, short, barrel-shaped tube on trees; usually the single door is hinged below. The fangs embedded in the back of the door hold it shut. Usually, these burrows have only one door but, as in ornate brush-footed trapdoor spiders (*Sason*), *Poecilomigas abrahamsi* has a door at each end of its burrow, making it hard to imagine how it keeps both safely closed.

WAFER-LID TRAPDOOR SPIDERS

The genus *Fufius* includes 13 species from Central and South America. It is unusual among mygalomorphs in that the transverse fovea (the cephalic groove marking the attachment of the sucking stomach) curves backward, not forward or straight as usual. The association of this and other African and South American genera with the original genus, *Cyrtauchenius*, on which the family is based is contentious—probably largely because specimens of only one species of that genus have been found since 1892. Many mygalomorphs make burrows with or without doors. Some make extensive webs, whereas a few live freely in the soil and leaf litter, and may themselves be a trap. However, the earliest form of web made by these spiders is a filmy set of tubes under logs and rocks.

Deep burrows in soils are often made by spiders in dry environments, allowing the spider to manage humidity, and to a lesser extent, heat. When such management is not needed, e.g., in non-flooded Amazonian rainforests, evolutionary pressure has evidently forced little change in burrow structure; such filmy webs constructed in protected locations are still all that is needed in several families. It is this kind of burrow that is made by at least some *Fufius* in rainforests.

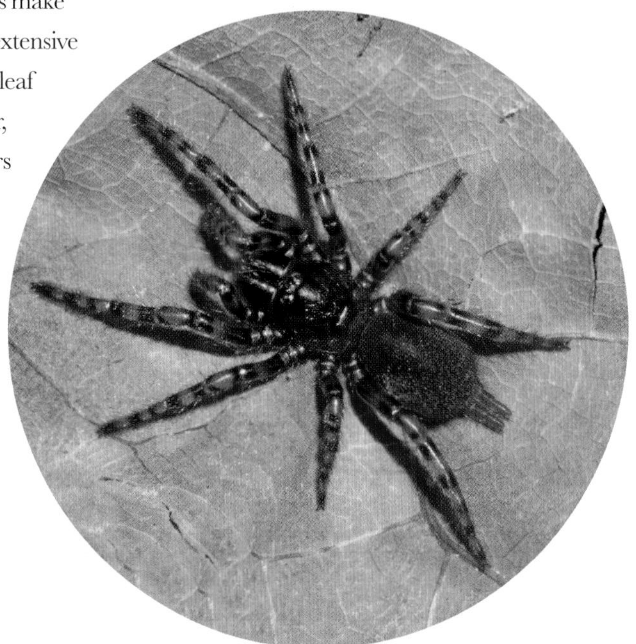

RIGHT | A banded-legged female *Fufius* species, showing the characters that challenge its taxonomic placement, including the long spinnerets, the similarly sized front and back legs, and the backward-curved transverse groove on its head.

GENUS
Fufius

DISTRIBUTION
Central America, Brazil, Trinidad, Colombia, Ecuador, Bolivia

HABITAT
Forest dwellers, making nests of silken tubes under logs and rocks

CHARACTERISTICS
- Three claws, with two rows of teeth on upper claws
- Rake-like spines (rastellum) on chelicerae
- Arched head
- Eyes on a low mound, if any
- Mouthparts less rectangular than leg coxae

CTENIZIDAE: *CTENIZA*

EUROPEAN CORK-LID TRAPDOOR SPIDERS

Once the proud family of a very large (albeit confused) group of genera, Ctenizidae has been successively redefined by recent studies, which have reduced it to just three genera. Two are from Europe: *Cteniza* itself and *Cyrtocarenum* (found in Cyprus, Rhodes, and mainland Greece), plus the South African genus *Stasimopus*, which may well belong elsewhere. *Cteniza* now includes only four species, which are known from France (Corsica) and Italy (Sardinia and Sicily). Presumably the distribution of the genus reflects areas untouched by the last glaciation.

Like most mygalomorph spiders, *Cteniza* have highly localized distributions—reflecting the limited dispersion of the young, by foot on a rainy night, to only a short distance from the maternal burrow. This results in very clumped distributions, with smaller burrows densely grouped around that of the large female. The trapdoor closing the burrow is a thick, cork-like lid. As in mouse spiders (Actinopodidae), males have very long palps but no processes on the front legs. This suggests that they do not elicit the common response of female mygalomorphs (to rear up into the "attack" position). Instead, the males remain safely away from the female's fangs while the long palps reach under her from afar to mate.

ABOVE | An adult female Moggridge's Trapdoor Spider (*Cteniza moggridgei*), Savona province, Liguria, Italy. The strongly arched head and more lightly built legs help identify members of this family.

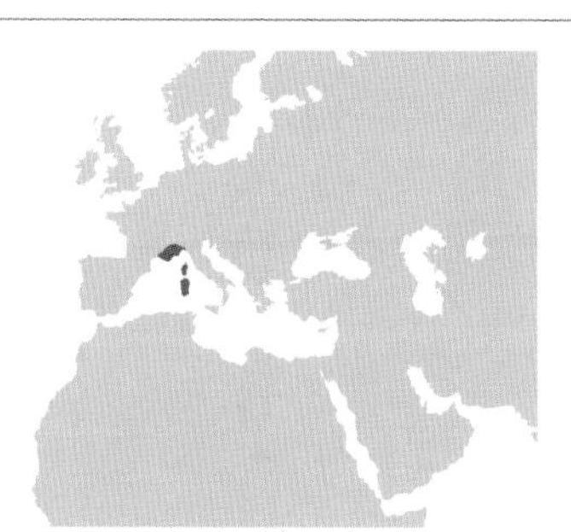

GENUS
Cteniza

DISTRIBUTION
Four species from France and Italy; one doubtful species from central Asia (unseen since 1875)

HABITAT
Closely grouped burrows with thick, cork-like trapdoors

CHARACTERISTICS
- Stocky, dark spiders with an arched head
- Eyes flat on the head
- Pronounced spiny "digging" rastellum on chelicerae
- Three claws
- Front two pairs of legs more lightly built than rear two pairs

HALONOPROCTIDAE: *CONOTHELE*

SADDLE-LEGGED TRAPDOOR SPIDERS

The remarkable and diverse trapdoor spider genus *Conothele* presently includes 26 species from Southeast Asia, New Guinea, and Australia. However, that is likely to be an underestimate. The genus may well be inseparable from a northern sister group, *Ummidia*, which currently includes 27 species from the New World, southern Eurasia, and northern Africa. Studies in eastern Australia alone indicate far more diversity than the three species now recognized there.

If all these species are indeed congeneric, this would be the only mygalomorph genus with a nearly global distribution. Certainly they are all united by the deep, saddle-shaped excavation found on tibiae III. This reinforcement allows the spiders to press sideways, thus anchoring themselves in their burrow and giving greater leverage to hold the door locked shut with their fangs embedded in the back of it. It may also serve to lock the spider and make both soil digging and wall packing easier.

Also of interest is that members of both genera are among the most unusual of

LEFT | A female *Conothele* from northern Queensland, Australia; note the smooth, shiny excavation on the tibia of the third pair of legs.

GENUS
Conothele

DISTRIBUTION
Southeast Asia, Australia

HABITAT
Burrows on flat ground, on the slopes of watercourses or as few cells attached to trees, in environments ranging from deserts to rainforests, mangrove mud, and trees overhanging the sea

CHARACTERISTICS
- Stocky, dark spiders with an arched head
- Eyes on a raised mound
- Pronounced spiny "digging" mound on chelicerae
- Saddle-shaped notch on tibia of third leg
- Three claws, males with clubbed tarsi

trapdoor spiders in that the young climb bushes to pay out silk and aerially disperse (balloon). This no doubt accounts for the presence of these spiders on many relatively recently formed Indo-Pacific islands; they have left the footprints of their dispersal across the shores of Pacific islands.

Burrows are usually quite short—only a few inches—and capped with a thick, well-sealing plug; they are found on flat ground, in the slopes of watercourses, or even as very short cells attached to trees. In Australia, the maximal extent of habitat divergence is evident, with burrows of different (mostly unnamed) species being found from extremely hot, dry, desolate deserts through to rainforest, mangrove mud, and even on trees overhanging the sea.

ABOVE | A female *Cyclocosmia ricketti*, showing the amazing hardened plate that blocks the burrow and protects the spider.

Another genus in this family, *Cyclocosmia*, is notable for the unusual shape of its abdomen, which ends abruptly in a hard disk with heavy sculpturing. When the animals retreat headfirst to the bottom of their burrow, the abdominal shield forms a plug (or false bottom). If a wasp or other predator enters the burrow, it encounters the hard plug and thinks it has found an empty burrow, rather than an occupied one. *Cyclocosmia* also has an interesting distribution, with two species in the USA, one in Mexico and Guatemala, and 10 in China and Southeast Asia.

EUCTENIZIDAE: *MYRMEKIAPHILA*

NORTH AMERICAN TRAPDOOR SPIDERS

Among its 12 species, all from the USA, *Myrmekiaphila* famously includes one named after Canadian rock singer Neil Young. The family has no remarkable features except for lacking the special modifications of the male idiopids, both on the palp and the clasping spurs on the first legs. As in a number of related families, the front pairs of legs are built more lightly than the back pairs; this is counterintuitive, as the front legs are used for burrowing and engaging males and predators.

The spiders make short burrows with thin doors, highly camouflaged by adhesive soil. Spiders that have thick plug doors doubly ensure the integrity of their entrance by inserting their fangs in the back of the door and locking it closed, presumably against both predators and water. The thin door of the *Myrmekiaphila* burrow cannot be so used, as it is too thin and would fold in, exposing the burrow. Unusually among mygalomorphs, these spiders hunt both day and night.

BELOW | A female *Myrmekiaphila comstocki*, showing an undifferentiated overall body shape and more lightly built front legs.

GENUS
Myrmekiaphila

DISTRIBUTION
USA

HABITAT
Short burrows with thin doors, camouflaged by soil

CHARACTERISTICS
- Rake-like spines (rastellum) on chelicerae
- Arched head, eyes on a low mound, if any
- Legs I and II more lightly built than III and IV
- Clasping spur in male not claw-like
- Spinnerets short and not evident from above

GIANT WESTERN AUSTRALIAN TRAPDOOR SPIDERS

LEFT | *Gaius villosus*—giant long-lived females present a fearsome appearance suggestive of the highly venomous funnel-webs but are only distantly related.

In remnant bushland patches of southwestern Australia, a trapdoor spider population attracted the attention of a young arachnologist, Barbara Main, more than 40 years ago. The very characteristic trapdoor and associated twigs radiating from the burrows of the spiders allowed Main to focus on the group. She numbered the burrows and observed the spiders as they grew and enlarged their burrows. Assuming that the spiders do not steal others' burrows, Main observed one spider in its burrow for 40 years; only in the 43rd year of study did a student note that the burrow door had become detached and the burrow was in disarray—presumably, the spider had finally succumbed. So died the oldest known living mygalomorph.

One of seven species in its genus, *Gaius villosus* is the largest species of the family worldwide, and is often mistaken for a highly venomous funnel-web (Atracidae) in Western Australia. Idiopidae occur from Africa and Madagascar though India, Southeast Asia, Australia, New Zealand, and South America. Remarkably, despite their high diversity in eastern Australia, they are absent from New Caledonia, once believed to be part of Australia. Most genera have a strong set of digging spines on the chelicerae and the eye group is highly modified, with the side eyes of the front row set well in front of the others. The defining features of the family are the strange male palp, which has soft cuticle extended like a tongue to the embolus tip, and the coupling spurs on the first leg of males. Most build trapdoors, but some make long tubes that extend up plant stems. The spiders hang from the tips of the stems, with their back legs in the burrow and their front legs poised to snatch passing moths.

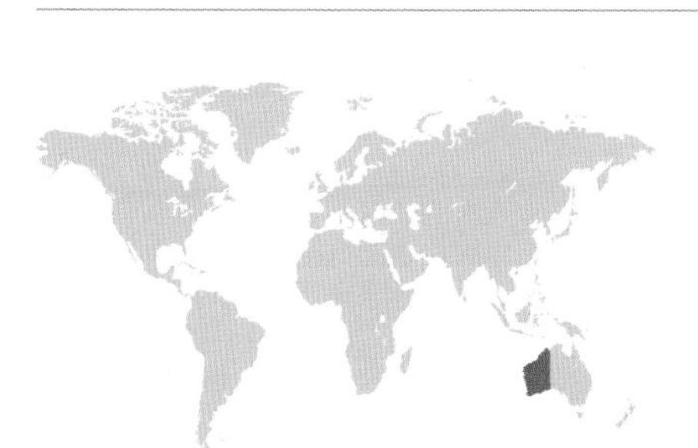

GENUS
Gaius

DISTRIBUTION
Western Australia

HABITAT
Burrows with distinctive trapdoors or long tubes running up plant stems

CHARACTERISTICS
- Short spinnerets, barely evident from above
- Large spiny digging process (rastellum) on the chelicerae
- Eight eyes in a trapezoidal group

ATRACIDAE: *ATRAX*

FUNNEL-WEB SPIDERS

It has been suggested that the more medically significant an organism is to humans, the greater taxonomic power it seems to assume. This may be apocryphal, but it has recently occurred: the genus *Atrax* and two other Australian funnel-web genera (*Hadronyche* and *Illawarra*), a total of 35 species, have been assigned family status, and they are without doubt among the most venomous spiders known. Prior to the development of antivenom in 1981, Australian funnel-web spiders were responsible for 13 deaths. In one rare case, a torso bite from one of these spiders was fatal in 15 minutes; bites to limbs

GENUS
Atrax

DISTRIBUTION
Eastern and southern Australia

HABITAT
Rainforest and open forest through to sclerophyllous heath; the tube may be well defined and funnel-like, but is usually closed and camouflaged

CHARACTERISTICS
- Large, shiny black spiders, quick to rise to a defense position
- Four spinnerets, these usually quite short
- Densely clustered black studs around the mouthparts
- Three claws

can be treated quickly by a bandage compress that slows the progress of venom through the lymphatic system. Surviving victims recall that within minutes of a bite you know that it is very serious: systemic symptoms involving fibrillation of both heart and lungs precede edema of the lung.

These remarkable spiders, however, are preyed upon by large lizards, which—like most animals aside from primates, including humans—are unharmed by the venom. The spiders feed primarily on apparently harmless vegetarian millipedes, their venom presumably paralyzing the prey, which when attacked sprays a cyanide-based chemical. Settling Australians quickly took to the picturesque real estate on Sydney's "North Shore," where the spiders are also most common. Speckled numerously through steep garden rockeries, the spiders' lairs are deep and safe, unlike those of the human intruders.

The venom is now being investigated in pharmaceutical research: it may help to arrest cancerous tumors in Tasmanian Devils (*Sarcophilus harrisii*) and melanomas in humans. Only the Sydney Funnel-Web (*Atrax robustus*) has proven fatal to humans, but this may simply reflect the dense concentrations of spiders and humans in the city, with far more bites occurring than serious envenomations.

With the slightest disturbance, these spiders rise to their attack-defense position, a drop of venom poised at the end of each fang. When male spiders encounter females, both sexes take up this attack position. The male then mates by going beneath the female, holding her fangs precariously above his head by a specially modified hook on his second legs.

LEFT | A female Sydney Funnel-web (*Atrax robustus*)—males of this species have been responsible for 13 deaths.

RIGHT | A female of the less dangerous western species, *Hadronyche marracoonda*, showing the impressive chelicerae, balanced leg thicknesses, and a flash of its orange book lung cover.

HEXATHELIDAE: *PLESIOTHELE*

LAKE FENTON SPIDER

The genus *Plesiothele* contains only one species, *P. fentoni*, which continues to thrive under deep winter snow in the Central Highlands of Tasmania. These spiders are unusual among their family for building simple, lidless burrows in mossy banks in heath and scree. Of their six spinnerets, four are small, but the anterior lateral pair are larger than the posterior median pair—in fact, they are the largest anterior spinnerets of any known mygalomorph, an indication of their ancient origins. Very little is known of the species' biology. The more widespread relatives of *Plesiothele* are delicate spiders that build flimsy tubes under logs and rocks, with slightly funnel-shaped entrances; some occur on trees. In the northern part of their range in Australia, they occur only in rainforests, or at least montane forests. Species occurring further south occupy lowland forests, but also have flimsy webs. The family includes one species with the most primitive excretory system of any known spider, whereas another species has the least developed spinnerets.

Most mygalomorphs have only four spinnerets, and a few have only two. According to conventional wisdom, the original number was eight, as found today only in liphistiids. However, the recently discovered amber spider fossil, *Chimerarachne*, has only four spinnerets in a completely different configuration from that of all mygalomorphs, but very like that of liphistiids. In a few of the more ancestral mygalomorph families, such as the Hexathelidae, two pairs of spinnerets are quite short. In life, the much longer back pair are long enough to be seen behind the spider, working like fingers to wave out the silk.

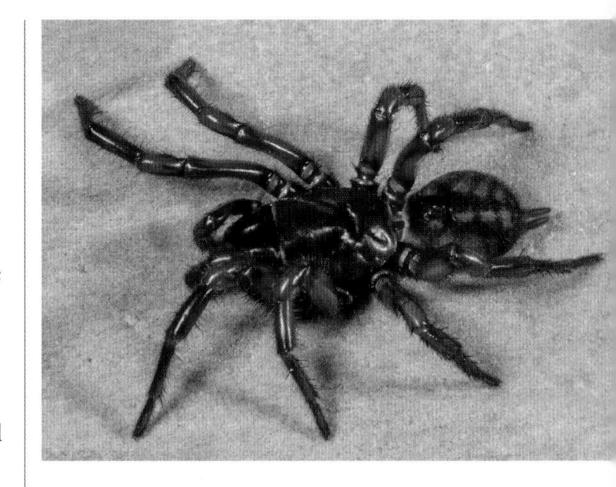

ABOVE | A female *Plesiothele fentoni*, with the largest anterior lateral pair of spinnerets, is one of the least evolved mygalomorphs. The ornate abdomen pattern is often not seen as the spiders live in silken tubes in alpine areas, often under snow.

Despite their phylogenetic separation from the highly venomous Atracidae, some of the very large black hexathelids in Tasmania readily bite—but apparently produce only minor, local reactions.

GENUS
Plesiothele

DISTRIBUTION
Tasmania

HABITAT
Flimsy tubes under logs and rocks or on trees; more rarely in burrows on mossy banks of scree and heathland. From rainforests to montane forests to lowland forests, and further south

CHARACTERISTICS
- Ornate, lightly built spiders
- Six spinnerets, two of which are long
- Densely clustered black cusps around the mouthparts
- Three tarsal claws

DIPLURIDAE: *CETHEGUS*

CURTAIN-WEB SPIDERS

Cethegus includes 12 species that occur from the hottest, wettest parts of Australia to some of the driest. Throughout their range, they build extensive webs in grasses and at the bases of trees, with several entrances; the spiders live in shallow burrows in the soil. This kind of web is built by many other members of the family; the South American relatives have absurdly long spinnerets that comb out swathes of silk as they walk. How they manage the extremely long spinnerets remains a puzzle, but on capturing prey, they rotate around the insect, wrapping it with silk as they go.

Although the members of *Cethegus* have only one row of teeth on the upper (paired) claws, the very long-spinnereted genera in the Amazon have two rows of teeth on those claws; some even have a sound-making lyra on the interfaces of the mouthparts. Judging by the number of snail shells in their webs, *Cethegus* and a related genera in moist parts of eastern Australia have mastered the slimy art of eating snails. The spiders' fangs stab the snail in the extended soft foot and relax it, preventing it from retracting inaccessibly into the shell, the usual reaction to an attack.

LEFT | *Cethegus ischnotheloides*—the very long spinnerets make expansive webs, which lead to underground tunnels.

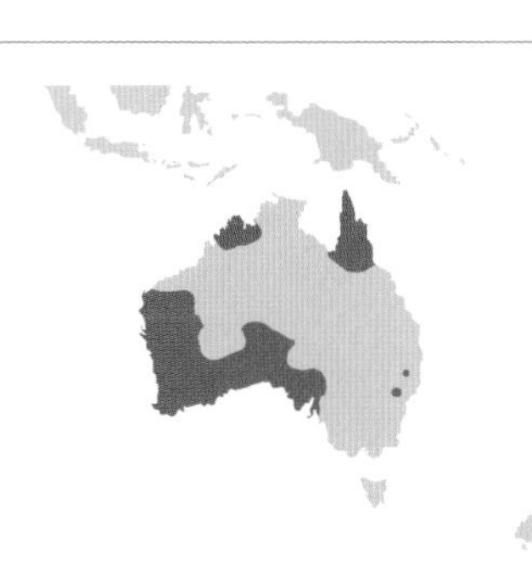

GENUS
Cethegus

DISTRIBUTION
Australia

HABITAT
Shallow burrows in soil, near grasses and trees

CHARACTERISTICS
- Four spinnerets, two of which are very long
- No cusps around the mouthparts, but with a rasp-like serrula
- Three tarsal claws

MACROTHELIDAE: *MACROTHELE*

NORTHERN TUNNEL SPIDERS

The sole genus in the family Macrothelidae, *Macrothele* occurs widely from the Congo, parts of North Africa, and Spain, through to China (although it is absent in the Middle East and similar adjacent drier areas). Until recently, *Macrothele* was considered closely related to the New Zealand tunnel spiders, the Porrhothelidae, and also to the highly venomous Australian funnel-web spiders, the Atracidae. Raventoxin—the venom of one species, *Macrothele raveni*—shares significant components with the venoms of at least some Atracidae. The only difference between Macrothelidae and Porrhothelidae lies in the shape of the mating spines on the first leg of the males, a character often of only genus-level significance in other mygalomorphs. In both families, the longer spinnerets sweep out curtains of silk as the spiders walk, allowing them to make extensive corridors in three-dimensional space.

LEFT | *Macrothele calpeiana*—the very long posterior lateral pair of spinnerets spin wide swathes of silk for its nest.

GENUS
Macrothele

DISTRIBUTION
Congo, parts of North Africa, and Spain through to China (although absent in the Middle East and similar adjacent drier areas)

HABITAT
Found in forests; make webs in banks and under logs and rocks

CHARACTERISTICS
- Large, shiny black spiders, quick to rise to a defense position
- Four spinnerets, two of which are very long
- Densely clustered black cusps around the mouthparts
- Three claws

PORRHOTHELIDAE: *PORRHOTHELE*

NEW ZEALAND TUNNEL SPIDERS

Until recently, *Porrhothele* was considered closely related to the northern tunnel spiders, *Macrothele*, and also to the highly venomous Australian funnel-web spiders, Atracidae. The only difference between Macrothelidae and Porrhothelidae lies in the shape of the mating spurs (spines) on the first leg of the males—a character often of only genus-level significance, but of great significance to the male and the species in general. The mating spurs are all that lies between the male and oblivion. In the mating prelude, he greets the female, who rises quickly to the attack position, her head and front legs raised and fangs open. He gingerly moves under the female and, with his mating spurs, immobilizes the fangs that are poised downward at his head and brain. The spurs take on a diverse set of shapes across the Mygalomorphae, but all serve much the same purpose. Sometimes an additional spur occurs on the second leg, which in the Dipluridae is used to spread and hold apart the female's legs, facilitating the safe transfer of sperm.

Porrhothele, like some Australian curtain-web spiders (Dipluridae), feed on snails. By biting deeply into the snail's foot and not releasing, despite being almost smothered in mucus, the spider can kill and eat the snail.

ABOVE | A female *Porrhothele* from New Zealand; these animals resemble *Macrothele* in having long posterior lateral spinnerets that produce swathes of silk.

GENUS
Porrhothele

DISTRIBUTION
New Zealand

HABITAT
Found in forests; make webs in banks and under logs and rocks

CHARACTERISTICS

- Large, shiny red-brown or black spiders, quick to rise to a defense position
- Four spinnerets, two of which are very long
- Densely clustered black cusps around the mouthparts
- Three claws

MICROSTIGMATIDAE: *PSEUDONEMESIA*

MICRO-MYGALOMORPH SPIDERS

The family Microstigmatidae currently includes only seven genera and 17 species, found from Panama to Argentina and in South Africa; many are notable for their tiny size. *Micromygale diblemma*, known only from Panama, is the smallest known mygalomorph; adult males are just 0.03 inches (0.75 mm) long; the species also (as its specific name indicates) has only two eyes.

The tiny size of these spiders mitigates against them being taken knowingly by hand in rainforest searches. Most specimens are taken mechanically, in Berlese funnels, using heat to extract the animals from the moist leaf litter in which they live. Far more species probably remain to be found.

These tiny spiders seem analogous to a type of salamander, the axolotl, i.e., they have retained

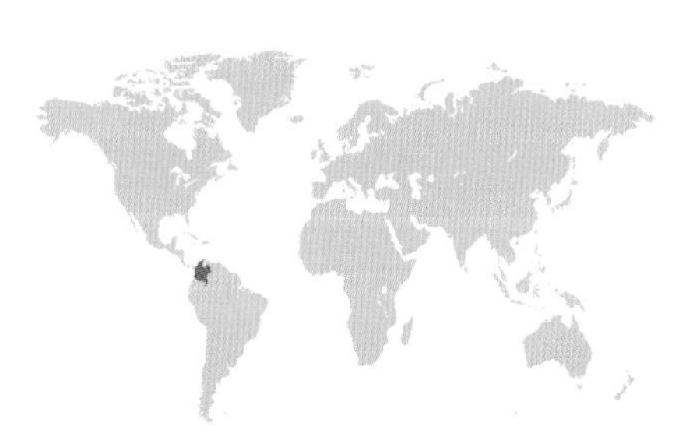

GENUS
Pseudonemesia

DISTRIBUTION
Venezuela and Colombia

HABITAT
Rainforests (among fallen leaves on forest floor)

CHARACTERISTICS
- Book lung openings small and round
- Scaly leg cuticle
- Strange modifications of serrula

"larval" or at least early developmental characters (small and round book lung openings) that no longer occur in most adult mygalomorphs. The family encompasses the full range of spinneret number, from the least developed (six), through four (the most common mygalomorph state), and then even to merely two spinnerets, a condition rarely seen. In addition, a rasp-like structure (serrula) on the mouthparts, normally composed of a group of small triangular spines, shows a variety of spine shapes, include one resembling a human hand. The spiders' small size presumably also reduces the pressure on the breathing organs (book lungs), and the normally long, ovoid slits are instead uniquely reduced to small round holes. The name *Pseudonemesia* means "false *Nemesia*"—the spiders were first associated with the family Nemesiidae, and they may well belong there or be closely allied with it.

There is one other group of mygalomorphs that are similarly tiny in size. The micro-curtainweb spiders of the genus *Masteria* run through the spaces among fallen leaves in the rainforest. These small spiders are currently placed in the family Dipluridae, but present characters that make them out of place there. They occur across the tropical Pacific from the Marshall Islands, northern Australia, New Guinea, and New Caledonia, through Fiji to South America, where they attain maximum diversity. Their island distribution raises the issue of whether they are recent colonists, but an Australian fossil male *Edwa maryae* from the late Triassic (220 million years ago) looks remarkably like its living relatives.

The first legs of males have a pair of specialized spines that presumably lock the female's fangs safely open and well above the male's head. The coupling spine shapes are clearly very efficient, as they have apparently evolved several times—in unrelated trapdoor spiders (Idiopidae), tarantulas (Theraphosidae), and brush-footed trapdoors (Barychelidae), as well as in *Masteria*.

LEFT | A male *Xenonemesia platensis*, showing the dull matte finish of the cuticle reflecting the scaly or roughened surface.

RIGHT | A female *Masteria toddae*. This tiny rainforest species has reduced eyes and runs through leaves on the forest floor. The spinnerets and similarly sized legs are clearly seen in this individual.

NEMESIIDAE: *XAMIATUS*

GIANT BEARDED TUBE SPIDER

The largest members of the Nemesiidae belong to *Xamiatus*, a genus that includes five species and is found in the rainforests of eastern Australia. Their size vies with that of the Australian tarantulas, which occur in nearby forests. When first discovered, these spiders were thought to be a remarkable, new, dark species of tarantula, but the reality was far more exciting. Their finely sculptured, bumpy cuticle and a special sensory lobe on the end of the tarsus hinted at relationships with a smaller, more ornate genus that makes tubes in the soil. *Xamiatus* builds a tube that belies its

LEFT | Members of this handsome species, *Xamiatus kia*, live in New South Wales, Australia, and construct a sinuous burrow, but deposit very little silk on the burrow's walls and entrance.

GENUS
Xamiatus

DISTRIBUTION
Eastern Australia

HABITAT
Found along creekside embankments in rainforests and wet sclerophyll vegetation. Unusually for the family, the burrow entrance, at least externally, lacks silk, and is not circular but irregularly oval

CHARACTERISTICS

- Three claws, the two upper paired claws with two rows of teeth
- Very large, dark spiders
- Fringe of silver hairs on inner face of first segment of palp and first leg

ABOVE | A male of the giant *Xamiatus magnificus*, showing the white hairs inside the front legs; the roughened cuticle of the legs and head create a matte appearance.

silk-making skills: the entrance is not circular, but asymmetrical, and there is little or no silk evident at the entrance or down the shaft. Almost without exception, when excavated—however carefully—these large spiders are hostile, lunging repeated at any disturbance, a behavior oddly seen in a number of Australian mygalomorphs. During these displays, the characteristic flashes of silver hairs along the sides of the bases of the front legs are even more impressive. Very little is known of the spiders' behavior and, no doubt because of their size, they are highly prized by the pet trade.

Other nemesiids occur on all continents except Antarctica. Some make simple tubes of silk under rocks and logs, whereas many make burrows without doors. Some of the latter have a Y-shaped entrance, with one arm of the "Y" set just below the ground and used primarily as an escape route.

PARATROPIDIDAE: *PARATROPIS*

ENCRUSTED TARANTULAS

These are among the most cryptic of tarantula-like spiders because they themselves are the trap. With their soil-encrusted cuticle, they make no burrow but instead rest in the surface layers of the soil, looking for all the world like they are part of it. The normally sensitive, fine hairs on the legs are highly reduced, with some protected by spines over small saddles or grooves in the upper legs; no doubt these are all triggers for prey walking near or over the spider. Although *Paratropis* presently includes only seven species, this is known to be a vast underestimate—in part because special techniques are needed to discover these small spiders, especially when garishly large tarantula burrows beckon nearby.

The highly reduced body hair and mounded eye group of the Paratropididae made their relationships difficult to understand until a link

LEFT | Male *Paratropis* are less adapted to cryptic hunting than the females and less heavily encrusted with soil.

RIGHT | A female *Paratropis*, showing the remarkable encrusting of algae that further aids the camouflage.

GENUS
Paratropis

DISTRIBUTION
Central and South America

HABITAT
Rainforest; the spiders conceal themselves well in the soil as a "body trap"

CHARACTERISTICS
- Bald but spiny, soil-encrusted legs
- Eight eyes on a raised mound
- First leg of male with claw-like coupling spur

between the genus *Melloina* and the tarantulas was established that showed more conservative modifications. Like the tarantulas, *Melloina* had more hair and claw tufts. In some mygalomorph spider groups, the legs reflect evolutionary sequences, with the first leg being the most advanced and the fourth leg the least. This front–back trend is reflected in the number of claws in some Nemesiidae, where only two of the normal three claws are on the first pair of legs. In Paratropididae, the reverse occurs, with a third claw present on the first leg but not on the fourth.

Many paratropidids are most notable for having their bodies encrusted with mud and soil particles, providing effective camouflage. A variety of araneomorphs have similar habits, including some sicariids and thomisids, but this behavior seems to be unique among mygalomorphs.

BARYCHELIDAE: *SASON*

ORNATE BRUSH-FOOTED TRAPDOOR SPIDERS

These little trapdoor spiders create their burrows in the trunk and branches of trees, with callouses, deep grooves, and lichens on the bark adding to the camouflage. There can be as many as six pairs of tiny doors, each pair back-to-back, in an area half the size of a dollar bill. The doors, each just a straw's diameter wide, close a tiny U-shaped tube in which the spiders hide.

Sason colemani is one of nine described species of tiny brush-footed trapdoor spiders of the genus *Sason* that occur on islands from the Seychelles in the western Indian Ocean through at least to northeastern Australia. These are beautifully ornate, tiny, rainforest-inhabiting species, whose relationships continue to confound us. They occur in lowland areas near coasts and have

RIGHT | A female *Idioctis littoralis*, whose watertight burrow is sealed tight as the tide rises each day. The very short spinnerets that weave the water-repellent silk are tucked almost unseen below the abdomen.

LEFT | A female *Sason colemani*—a tiny, fast, banded-legged trapdoor spider.

GENUS
Sason

DISTRIBUTION
Indian region, Southeast Asia, Pacific islands, Australia

HABITAT
Burrow on the trunk and branches of trees, using grooves of bark and lichens as camouflage; found from rainforests to lowland coastal and intertidal regions, and mangrove forests

CHARACTERISTICS
- Small, boldly marked spiders
- Claw tufts and short spinnerets
- Burrow on trees with two doors back-to-back

undoubtedly traveled widely on floating palms, leaving their footprints in the sands as they went. Nevertheless, each of the nine recognized species is quite different and distinct.

The movement from the land back to the sea, from which all animals once came, fascinates us all. Members of *Idioctis*, another barychelid genus, have become intertidal, using burrows lined with waterproof silk and capped with a floppy silk-lined door, adorned with algae. The spiders hold this closed with their front legs, and maybe also their fangs. One of the nine species, *I. yerlata*, builds its tubes in small holes in mangrove trees and coral near Australia's Great Barrier Reef. The burrows are submerged at half-tide. Like many trapdoor spiders, these are nocturnal hunters—not only must they wait for darkness before they can hunt, but they can also hunt only up until the half-tide. Like *Sason*, these spiders occur from the Seychelles through Singapore to eastern Australia, but also eastward to Fiji and Samoa.

The species all look quite similar, but live in dramatically contrasting places in the intertidal zone. In Singapore, *I. litoralis* makes its burrow in the large triangular mounds, about waist high, made by mud lobsters—evidently an easily raided substrate, but the spiders flourish regardless. In southern New Caledonia, *I. ferrophila* occupies relatively short tubes in large ironstone boulders, which resist even a chisel driven with a small sledgehammer. Significantly, it is now the species farthest from the equator, and no doubt takes advantage of the latent heat of the great boulders.

THERAPHOSIDAE: *BRACHYPELMA*

ORANGE-KNEED TARANTULAS

RIGHT | A strikingly marked female of the highly threatened Orange-kneed Tarantula (*Brachypelma smithi*). These colorful spiders are much sought after by the pet trade, and are now intensively reared in captivity for this purpose.

Once highly sought after for the early pet trade, these large, ornate, but quite docile tarantulas are most common in Mexico, where they have been heavily exploited—to the extent that they are one of the few spider groups whose international movement is noted by governmental authorities. They are listed as part of the US Endangered Species Act and as Threatened Species by the IUCN on their Red List. As with many of the South American tarantulas, *Brachypelma* quickly brush the sides and top of the abdomen with their back legs when disturbed, producing a "cloud" of tiny hairs.

ABOVE | A female Orange-kneed Tarantula (*Brachypelma smithi*)—this species is large and docile. The extent of color on the legs varies among species.

GENUS
Brachypelma

DISTRIBUTION
North and Central America

HABITAT
In forests, the spiders build open holes with deep burrows

CHARACTERISTICS

- Two claws flanked by dense claw tufts
- Urticating (itching) hairs on abdomen
- Colorful bands on legs and abdomen

Each of these has multiple barbs designed to deter an offending mammal predator.

Theraphosidae includes the largest spiders, by weight, if not by leg span. As an indication of size, one giant *Theraphosa blondi* in the American Museum of Natural History has been surgically cut in half lengthwise. The accompanying label states that the only tool available to stop this monster was a machete; the cut was clean, such was the spider's size. These large spiders have been filmed resting on trees at night, and even taking bats on the wing.

Australasian tarantulas are not so docile. They attack quickly, often with a fine hiss or rattle elicited by the movement of stiff brushes on the mouthparts and pins on the opposing lower outer face of the chelicerae. In some species, the audible sound is strongly amplified by what is effectively a boom box, formed by a dense brush of hairs adjacent to the brush.

Tarantulas also possess spectacular dense pads of iridescent hairs (scopulae) on the tarsi and metatarsi of their legs. The hairs effectively hold prey with smooth cuticle. However, when held in glass cases in captivity, the adhesion so formed, at least against the corners, has been sufficient to enable the spider to climb straight up the glass, push up the apparently substantial lid, and escape.

In Australia, tarantulas have been seen seizing, killing, and consuming the Australian Cane Toad (*Rhinella marina*). Being effective against vertebrates, tarantula venom is of special interest, and the venom of at least one species, *Phlogius crassipes*, is under research for its ability to bind to breast cancer cells.

Tarantulas occur throughout the world, in habitats ranging from the Valley of the Cross in Jerusalem, Israel, to icy mountaintops in South America.

ARANEOMORPHAE

The infraorder Araneomorphae contains most of the spider families, genera, and species—all spiders other than the liphistiids (Mesothelae) and mygalomorphs. Araneomorph (or true) spiders differ from liphistiids and mygalomorphs in the way their chelicerae move—the chelicerae are "diaxial" (moving primarily from side to side, with the fangs usually oriented transversely), rather than paraxial (moving primarily up and down, with the fangs oriented longitudinally)—see pages 20–21. But there are other differences as well, particularly in the spinnerets. As in *Liphistius*, araneomorph spider embryos have four pairs of spinnerets (anterior laterals and medians, and posterior laterals and medians).

In liphistiids, the posterior lateral spinnerets are multisegmented, but in mygalomorphs these spinnerets have just three or four segments, and in araneomorphs they consist of only one or two segments. Similarly, in liphistiids the adults either retain all four pairs of spinnerets, or have the two posterior median spinnerets fused to each other. In mygalomorphs, there is no trace of the anterior median pair, even in embryos.

LEFT | An araneomorph spider, *Dysdera crocata*. This Mediterranean species is synanthropic (lives in and around buildings) and has been widely introduced elsewhere (e.g., in Australia, Hawaii, the USA, and Chile). It is a specialist feeder on woodlice.

CRIBELLATE AND ECRIBELLATE ARANEOMORPH

Ventral view of the posterior end of the abdomen of a cribellate and ecribellate araneomorph. The top artwork shows an entire and divided cribellum—a transverse plate in front of the spinnerets bearing special ribbed spigots that produce woolly silk. The bottom artwork shows an ecribellate araneomorph with a colulus—a short protuberance in front of the spinnerets that is a modification of the cribellum. The colulus of ecribellate araneomorphs bears no spigots and does not produce silk.

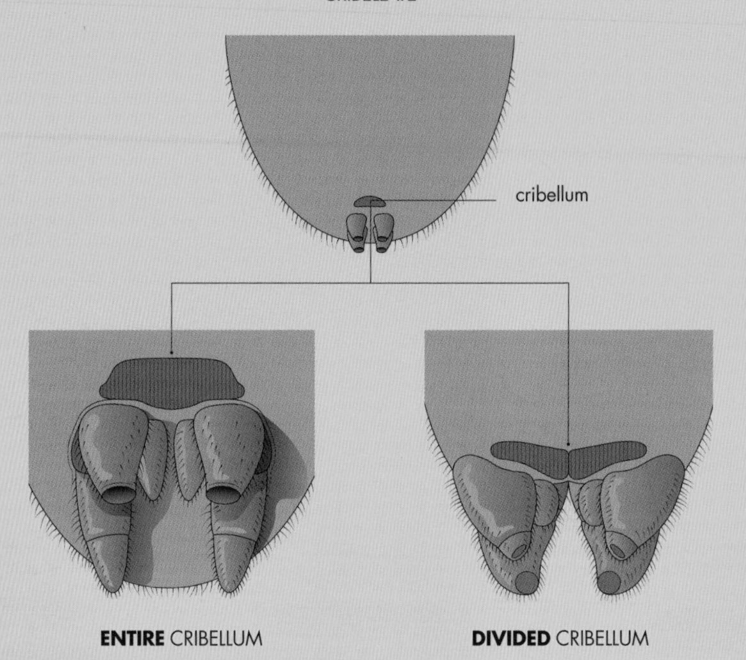

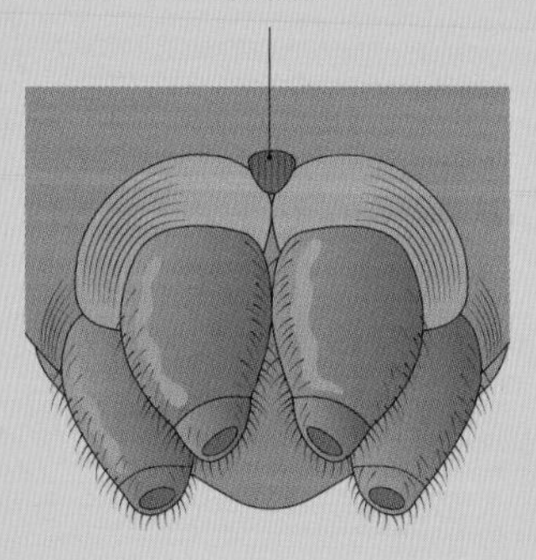

LEFT | The ribbed spigots found on the cribellum of araneomorphs. They differ in shape from the various spigots on the other spinnerets, which produce sticky, rather than woolly, silk.

In araneomorphs, the anterior median pair is present in embryos, but fuses together during development. If the fused pair remains functional in the newly hatched spiderlings and adults, it forms a cribellum—a transverse plate bearing special, ribbed spigots. Cribellate silk is "woolly." Its adhesiveness is both mechanical and due to Van der Waals adhesion forces (a property of materials in very close contact), as the silk consists of hundreds of tiny fibers loosely wrapped around a few thicker, core fibers. To pull silk from the cribellum, spiders use a series of specialized hairs on the metatarsi of their hind legs, the calamistrum.

Even in those araneomorphs that have a functional cribellum and calamistrum, males typically lose those functions in their final molt, when they become adult. Once adult, they no longer build webs to capture their food, but rather wander in search of females. But only a minority of araneomorphs retain a functional cribellum. In most, the cribellum is represented in adults by a much smaller, usually triangular, lobe called a colulus, which bears no spigots and produces no silk. These animals (the ecribellate araneomorphs) rely on a variety of strategies. Some make viscid, sticky silk, produced from spigots on the posterior lateral spinnerets, others rely on mechanical entanglement by dry fibers, and many others hunt their prey actively and do not rely on prey-catching webs at all.

The more basal groups of araneomorphs have relatively simple ("haplogyne") genitalia, in which the eggs, when being laid, are fertilized with sperm that travel from the spermathecae, where they are stored in the female, through the same duct that the male uses during mating to introduce the sperm into the female's body (see pages 104–105). That does not make the haplogyne araneomorphs a group, however; liphistiids and mygalomorphs also have haplogyne female genitalia, so it is the primitive condition for spiders in general.

HYPOCHILIDAE: *HYPOCHILUS*

LAMPSHADE SPIDERS

The 10 known species of lampshade spider are endemic to small, widely separated montane regions of the USA. They construct unique webs in the shape of a lampshade. The narrow circular end of the "lampshade" is attached to a vertical rock surface below an overhang. The spider rests in the middle of the web during the day, totally exposed yet hardly visible thanks to its cryptic coloration—very similar to the rock texture or to the lichens and mosses that cover the rock surface. It is hence easier to spot the webs than the spiders.

During the night, spiders can be seen capturing insects entangled in the web. They comb their cribellate sticky silk using a stereotyped movement of leg IV supported by the opposing leg III—a behavior shared only with filistatids.

There is only one other genus in the family Hypochilidae, *Ectatosticta*, with two species living in China. Along with a handful of araneomorph spiders that retain all four book lungs, hypochilids are considered "living fossils."

RIGHT | A female *Hypochilus pococki* at Great Smoky Mountains National Park, North Carolina, USA. During the day the spider rests flat on the rock surface, in the middle of the web and facing downward.

LEFT | A male *Hypochilus petrunkevitchi* perfectly camouflaged over a rock at Sequoia National Park, California, USA. The males have very long, thin legs that reach up to 6 inches (15 cm) in length.

GENUS
Hypochilus

DISTRIBUTION
USA

HABITAT
Montane, with webs fixed to overhanging rocks

CHARACTERISTICS
- Four book lungs
- Cribellate
- Makes a lampshade web

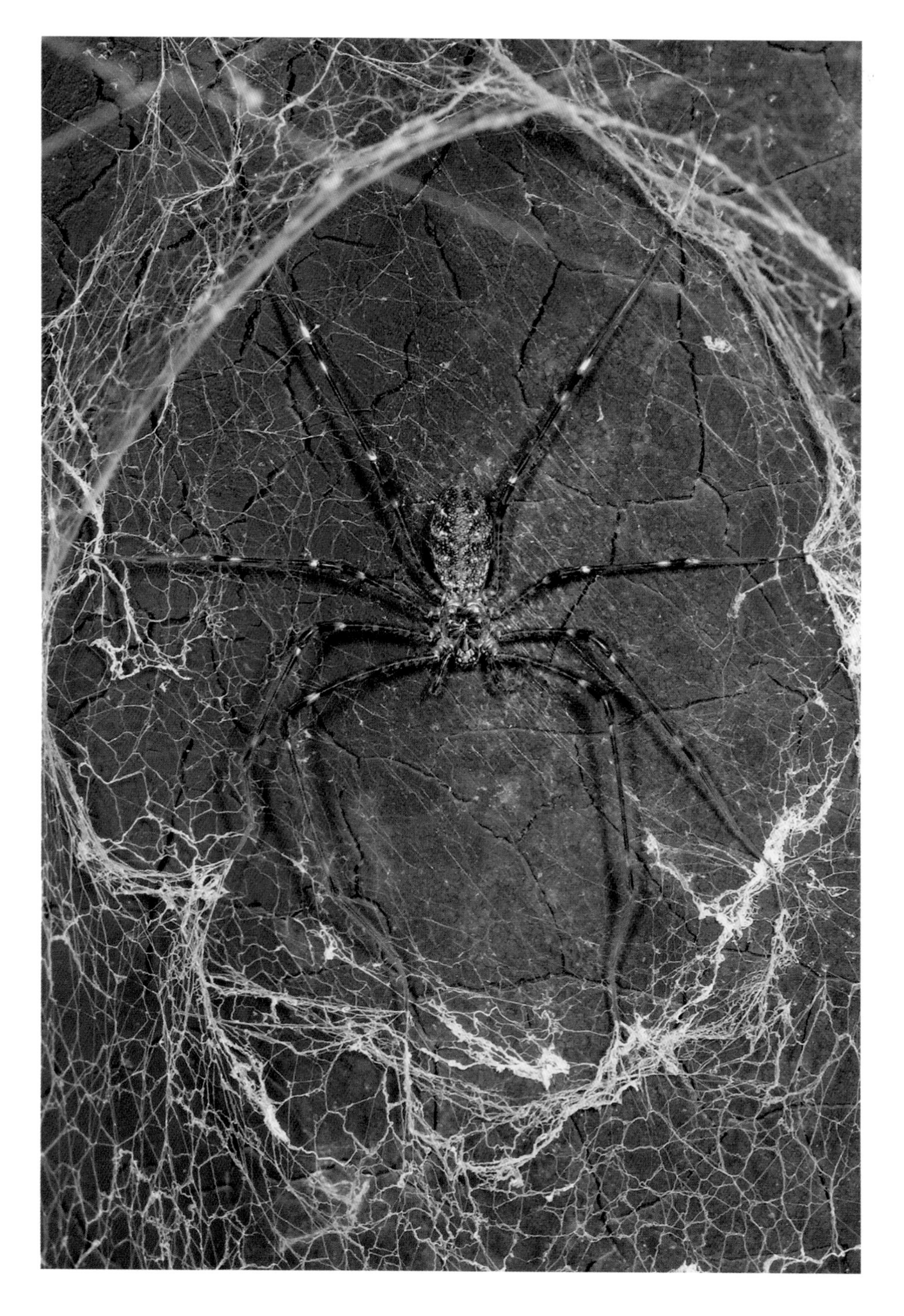

AUSTROCHILIDAE: *HICKMANIA*

TASMANIAN CAVE SPIDER

The Tasmanian Cave Spider (*Hickmania troglodytes*), the only species in its genus, is another "living fossil" araneomorph spider that retains the primitive configuration of four book lungs. It builds large, horizontal cribellate webs in caves or in rotten, hollow tree trunks in forests, which can reach more than 6 feet (1.80 m) in length. The spiders are nocturnal and extremely shy, hiding or becoming motionless upon the slightest disturbance.

The males are very long-legged, and the metatarsi of the second pair of legs have a pronounced curve. During mating, the male uses these metatarsi to grasp the female's chelicerae, keeping her fangs cautiously at distance. Females make a drop-shaped egg sac of very hard white silk, and hang it deep inside the funnel-like retreat of the web. Females living in the forest camouflage the egg sac with a layer of litter particles. The young spiderlings emerge after eight to ten months—a much longer gestation period than most spiders. They appear to be extremely long lived, with life spans that probably reach several decades. A feature movie, *Sixteen Legs*, has been devoted to this spider species.

Hickmania troglodytes is a representative of a very old lineage. The species is probably most closely related to the South American *Austrochilus* and *Thaida*, which are endemic to the temperate austral forests of Chile and Argentina.

LEFT | A male *Hickmania troglodytes* in Marakoopa Cave, Tasmania. During mating, the males use the long, curved second legs to keep the female chelicerae at a safe distance.

RIGHT | A male *Hickmania troglodytes* at King Solomon's Caves, Tasmania. The four book lungs are evident as white patches on the abdomen.

GENUS
Hickmania

DISTRIBUTION
Tasmania

HABITAT
Caves or hollow tree trunks in forests

CHARACTERISTICS

- Four book lungs
- Cribellate
- Males have a curved metatarsus II

OTWAY ODD-CLAWED SPIDERS

RIGHT | A penultimate male of *Progradungula otwayensis* holding an intact catching ladder. The spiders employ a highly stereotyped sequence to spin the ladder. The two vertical lines of the ladder are made of cribellate silk that is laid directly, without combing. To date, this is the only known case of cribellate silk spun without the intervention of the calamistrum.

LEFT | A *Progradungula otwayensis* in hunting position, holding a web that is partially worn out. The bluish cribellate silk is adhesive and impressively extensible, stretching up to 15 times its original length.

GENUS
Progradungula

DISTRIBUTION
Great Otway National Park, Victoria, Australia

HABITAT
Hollow trees in Myrtle Beech or Mountain Ash trees from old growth forest

CHARACTERISTICS
- Four book lungs
- Asymmetric claws on legs I and II, the superior internal claw enormous and raptorial
- Spins a highly stereotyped ladder web of cribellate silk

The Otway Odd-clawed Spider (*Progradungula otwayensis*) occurs only in a small area of old-growth forest in southern Victoria, Australia. The animals spend the day in hollow trees in Myrtle Beech (*Nothofagus cunninghamii*) or Mountain Ash (*Eucalyptus regnans*) trees. At sunset, they walk along a silken line to a hunting spot up to 10 feet (3 m) away from their retreat, where they construct a highly regular catching ladder made of bluish cribellate silk. Here the spiders position themselves, tensing the web and extending the two front pairs of legs as they wait for prey to pass nearby. One of the tarsal claws on these legs is much larger than the other; it is raptorial, like a folding knife. Similar predatory behavior has been seen in a close relative, the Carrai cave spider (*P. carraiensis*), of New South Wales, Australia. When an insect walks within the range of the spider's front legs, it suddenly thrusts and clasps the prey with the claws, pushing it into the ladder of sticky silk. The prey is enmeshed and then quickly bitten and subdued.

All the members of the small family Gradungulidae occur in Australia and New Zealand. Most species do not build webs and are active nocturnal hunters, but still have unequally sized tarsal claws. Because they retain four book lungs, as in liphistiids and mygalomorphs, gradungulids are considered "living fossils."

CREVICE SPIDERS

Kukulcania hibernalis is one of the spiders most commonly found inside buildings, from the southeastern USA to central Argentina. They belong to a desert-inhabiting lineage, and are thus preadapted to sedentary living inside buildings, with no access to water. There are several *Kukulcania* species endemic to North and Central America; in the wild they make their webs in rock crevices, soil burrows, under stones, or under logs. They build a funnel-like retreat for resting, from which sticky cribellate lines radiate. The webs are permanent, so the initially neat web receives layer on layer of new silk and becomes messy over time. The spiders card the cribellate silk in a peculiar way, as only hypochilids do; the cribellate silk is heavily folded and extends greatly when an entangled insect tries to get away. Unlike other araneomorph spiders, filistatid females continue to molt after reaching maturity, and can live for many years.

Filistatids are a puzzling lineage with a unique appearance and a mixture of very primitive and uniquely derived characters. Their closest relatives are probably the hypochilids.

LEFT | A male *Kukulcania hurca* wandering at night in desert wash habitat in San Diego, California. The males are much paler than the females and have very long legs with flexible tips. They wander at night in search of females.

RIGHT | A female *Kukulcania hibernalis* combing the cribellate silk. The fourth leg vibrates rapidly and is supported with the opposed third leg. The silk is folded in a zigzag and can be extended over 25 times its original length.

GENUS
Kukulcania

DISTRIBUTION
Probably native to southeastern USA and Mexico; now introduced to most of Central and South America

HABITAT
Buildings; in the wild in rock crevices and soil burrows, under stones or logs

CHARACTERISTICS
- Cribellate
- Eight eyes clustered on a mound
- Females dark brown; males pale, with very long legs
- Spin heavily folded cribellate silk

SICARIIDAE: *LOXOSCELES*

RECLUSE SPIDERS

Loxosceles spiders are famous for their strong venom, which causes necrosis (tissue destruction around the site of the bite) in humans. Because several species are synanthropic (living in and around buildings), bite incidents occur with these spiders periodically, when a spider is being crushed and bites defensively. Molecular studies have shown that dispersal routes of the Mediterranean Brown Recluse Spider (*L. rufescens*) coincide with those of humans around 40,000 years ago.

In nature, members of this species are found mostly under rocks and in caves, but because of human transport, they are now widespread. For example, the species was first recorded in Western Australia in 1996, arriving in cargo from Singapore. Others of the 130-plus species described worldwide have smaller distribution ranges; several live in natural habitats, including caves, although only a few species show true cave adaptations such as eye reduction. They are

GENUS
Loxosceles

DISTRIBUTION
Tropical and subtropical regions, Mediterranean; some species introduced to USA, Mexico, Macaronesia, South Africa, Finland, India, China, Japan, Korea, Laos, Thailand, Australia, Hawaii

HABITAT
Various natural and man-made habitats: under stones, in holes in the ground, under tree bark, in caves, in crevices in and around buildings

CHARACTERISTICS
- Six eyes, arranged in three groups of two
- Venom causing necrosis in humans
- Spin flattened ribbons of silk using only their spinnerets

LEFT | A female *Loxosceles* from Belize, showing the irregular, loose web with its characteristic white cribellate threads.

ABOVE | An adult female *Loxosceles rufescens*, from a cave in Cambodia. In Laos and Cambodia the species was found exclusively in caves with human impact. The spiders were very calm, sitting in their webs on horizontal surfaces like the ground or rock projections.

sometimes called violin spiders, due to the characteristic coloration pattern on their carapace.

Whereas most spiders use their legs, or the substrate, to help pull silk from the spigots, *Loxosceles* can spin the flattened, woolly silk used in their webs with only the spinnerets themselves. The anterior lateral spinnerets are moved rapidly, up to 13 times a second, with the accumulating silk being held by the posterior spinnerets.

The other genera in the family are *Sicarius*, with 21 species in Central and South America, and the similar *Hexophthalma*, with eight species in southern Africa. These animals also produce medically significant venom, but with fewer bite incidents. Females are often found guarding their egg cases under rocks, but when disturbed their typical defensive behavior is to freeze, feigning death. In this mode, if one uses a pair of forceps to raise one of their legs in the air, they will even keep the leg held there, rather than show any movement at all. These animals are adapted to sandy habitats: the surfaces of their body and legs have modified hairs that help trap sand grains, providing them effective camouflage. In addition, they can rapidly burrow under loose sand to escape potential predators.

SCYTODIDAE: *SCYTODES*

SPITTING SPIDERS

Scytodids are unique spiders. They capture their prey by rapidly spitting a double zigzag of glue from their chelicerae, thus entangling the insect and fixing it to the substrate, while keeping a safe distance. The glue is produced by a special region of the massive venom glands that occupy most of their bulging carapace, and is ejected from the cheliceral fangs. The ejection speeds have been measured at more than 30 yards (28 m) per second. Once delivered, the glue contracts and further immobilizes the prey, which is afterwards bitten, and then consumed. These animals will attack and consume other spiders as well as insects.

Although a few species have synanthropic habits and are therefore very widespread, the genus is very diverse: more than 225 different species live in various parts of the world. They are nocturnal, and females carry their egg sacs with their chelicerae until the spiderlings emerge and disperse. Several scytodids make webs. Four similar genera are recorded from Asia.

LEFT | A spitting spider (*Scytodes*) at Durian Loop, Singapore. The females carry their egg sacs with the chelicerae until the spiderlings emerge. Spitting spiders have a distended carapace to make room for the enormous venom glands.

RIGHT | A female *Scytodes thoracica* from Austria. Spitting spiders are nocturnal and slow moving, except for the fast spitting of glue that they use to entangle prey from a safe distance. Several *Scytodes* species such as this one are common in houses and gardens.

GENUS
Scytodes

DISTRIBUTION
Cosmopolitan

HABITAT
All habitats, including in and around buildings

CHARACTERISTICS

- Six eyes in three pairs
- High, domed carapace
- Thin legs, the anterior two pairs with asymmetrical superior claws; one pair has a double row of teeth
- Spits zigzags of a special kind of glue from the chelicerae

PERIEGOPIDAE: *PERIEGOPS*

WIDE-CLAWED SPIDERS

Periegops is a small genus with three species, known from only a handful of localities in New Zealand and Queensland, Australia. They seem to require humid forests with deep leaf litter and well-drained soil. Little is known about their habits; they do not construct a web and are probably nocturnal hunters.

Periegops has long been a taxonomic mystery. Its species have many morphological characters linking them to drymusids and scytodid spitting spiders, despite their very different, stocky appearance and short legs. Molecular evidence suggests that they are the only representatives of a lineage of scytodoids that became geographically isolated long ago, and developed unique morphological specializations to a webless lifestyle.

BELOW | A female *Periegops suteri*. This species lives in a couple of localities near Christchurch, on the South Island of New Zealand.

GENUS
Periegops

DISTRIBUTION
Two species live in New Zealand, and a third is known from Queensland, Australia

HABITAT
Humid forests with deep leaf litter and well-drained soil

CHARACTERISTICS
- Six small eyes in three widely spaced pairs
- Haplogyne genital system
- Short legs, the anterior two pairs with asymmetrical superior claws; one claw of each pair has a double row of teeth

DRYMUSIDAE: *DRYMUSA*

FALSE VIOLIN SPIDERS

The small family Drymusidae includes spider species that build webs in dark, humid places under logs, rocks, and in caves. In tropical rainforests, they also make webs in spaces in the leaf litter. The web is an irregular sheet from which the spider hangs. Drymusids are placed in two genera: *Izithunzi*, from South Africa; and *Drymusa*, from South and Central America. Their common name refers to their resemblance to the venomous violin spiders of the South African genus *Hexophthalma*, but they lack the proteins responsible for the toxicity of the venom of those animals.

BELOW | A female *Drymusa* from Panama, carrying the egg sac with the chelicerae. These small, cryptic spiders make their sheet webs in the soil between fallen leaves in rainforests.

GENUS
Drymusa

DISTRIBUTION
From Cuba and Costa Rica south to Argentina and Chile

HABITAT
Tropical rainforests (among leaf litter); dark, humid locations under rocks or logs, or in caves

CHARACTERISTICS
- Six eyes in three pairs
- Flat carapace
- Haplogyne genital system
- Long legs, the anterior two pairs with asymmetrical superior claws; one claw of each pair has a double row of teeth

LEPTONETIDAE: *CALILEPTONETA*

MIDGET CAVE SPIDERS

Leptonetids are small spiders. They build their webs in dark, moist places such as caves, rotting logs, deep spaces in soil and leaf litter, and under stones. Many species live only in caves and have the typical morphology of life in total darkness: eye loss, elongated legs, and pale coloration. Unlike most spiders, their slender legs are weak and break between the patella and tibia; they also have a green or bluish shine, due to light interference from the microscopic sculpturing on the cuticle. Most leptonetids that retain eyes have them in a unique arrangement—a group of four anterior eyes and two set apart, well behind the others.

The evolutionary links of leptonetids have been puzzling, but recent genomic data suggest that they are an early branch on the araneomorph tree of life. In this new light, leptonetids seem to have been among the first lineages to acquire a type of silk

RIGHT | A *Leptoneta infuscata* from a cave in Banyuls-sur-Mer, France. Many leptonetids inhabit caves. The iridescent reflection of the legs of these minute spiders is produced by light interference from the microsculpture of the cuticle.

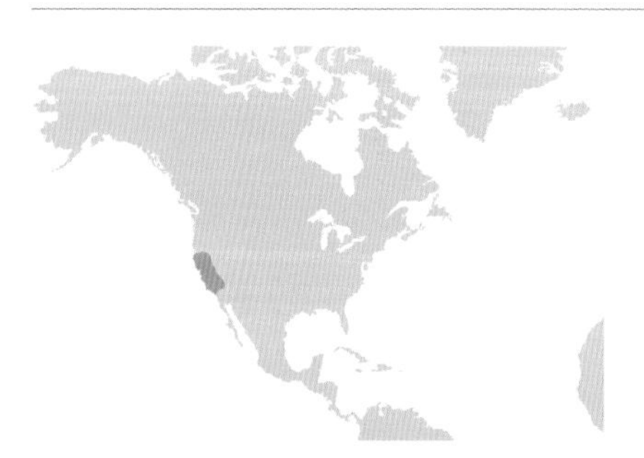

GENUS
Calileptoneta

DISTRIBUTION
California and Oregon, western USA

HABITAT
Dark, moist locations in caves, hollow logs, soil and leaf litter; under stones

CHARACTERISTICS
- Six eyes, four in an anterior group and two separated behind
- Three claws; leg autotomy (point of weakness) between the patella and tibia
- Ecribellate, with the spigots aligned on the posterior spinnerets

gland used exclusively for the construction of the egg sac—the cylindrical gland. With the exception of *Archoleptoneta*, all leptonetids are ecribellate. The nine known species of *Calileptoneta* make sheet webs in forests and caves in California and Oregon, USA. The male *C. helferi* has very long palps.

ABOVE | A male *Calileptoneta helferi* from Mendocino County, California, USA. This species makes delicate sheet webs under logs in humid forests.

TELEMIDAE: *TELEMA*

LONG-LEGGED CAVE SPIDERS

Telemids are minute spiders that live in dark, humid places, especially caves and interstices of the soil and leaf litter. Species of the genus *Usofila*, which is endemic to the USA, have six eyes, but many cave-dwelling telemids are eyeless. *Telema tenella* is one of those blind species; it lives in caves in the Pyrenees, at elevations above 3,000 feet (900 m). They construct a small sheet web.

The reproduction of these spiders is unique. The male produces a spermatophore—a cylindrical structure containing several tightly packed spermatozoa, which is inserted and stored in the female spermatheca. The female makes an egg sac with no more than three or four eggs, sometimes even only one; the spiderlings take 11 months to hatch, a very long time for a spider. They then take three years to reach maturity, with only one molt per year. Females have been reared for four years, and scientists believe they can reach 12 years of age—again very unusual for such a small spider. The slow development and long life span are probably related to the low temperature inside the caves they inhabit.

RIGHT | A female of an unidentified telemid, at Bukit Timah, Singapore. This species carries the egg sac with the spinnerets, but other telemids attach it to their web.

GENUS
Telema

DISTRIBUTION
Caves in the Pyrenees of France and Spain

HABITAT
Dark, humid locations, including caves; within soil and leaf litter

CHARACTERISTICS
- Minute eyeless spider
- Lives in caves
- Males transfer spermatozoa in a cylindrical package

MIDGET GROUND WEAVERS

The genus *Althepus* is endemic to Southeast Asia. It contains about 60 known species, with many more probably awaiting discovery. They make their webs in tree buttresses in the forest; many species live in caves. Ochyroceratids are small spiders that live in dark, humid habitats in the tropics all over the world. Most of the species are tiny, except for some Southeast Asian members of the subfamily Psilodercinae; these can resemble a pholcid in size and in having extremely long legs. Many ochyroceratids have characteristic purple or blue markings, and their legs have a green or blue shine. They make a horizontal sheet web, which on close inspection is seen to be made of layer over layer of highly ordered, parallel lines.

Species of the American genus *Ochyrocera* have a very narrow genital opening and produce very large eggs—which implies that the eggs must be fluid enough to squeeze through a small duct that is 40 times smaller than the deposited egg.

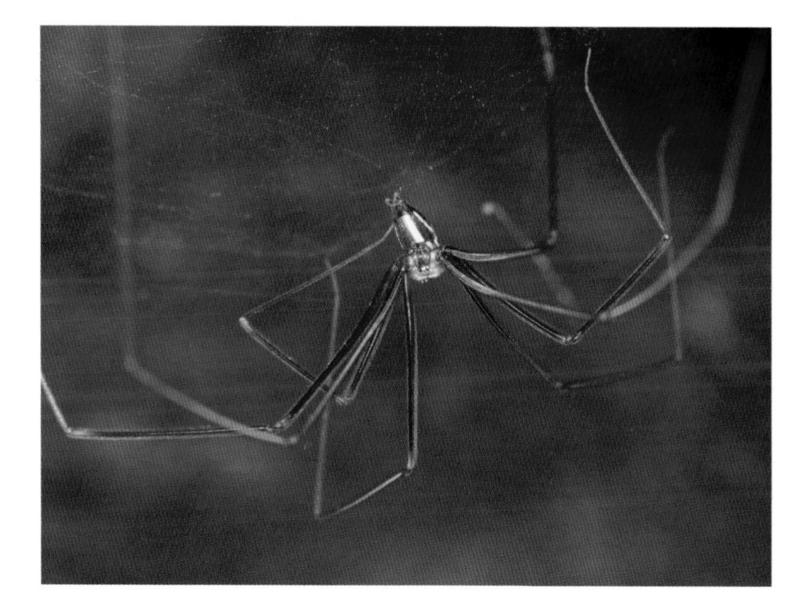

LEFT | A long-legged male of Ochyroceratidae subfamily Psilodercinae at Tham Kieo, Laos. The rainbow shine of the legs is caused by the microscopic sculpturing on the cuticle. The fine legs are extremely fragile and become detached easily.

GENUS
Althepus

DISTRIBUTION
Southeast Asia

HABITAT
Dark, humid forests and caves in the tropics

CHARACTERISTICS
- Six eyes
- Haplogyne genital system
- Spigots of posterior spinnerets in a line
- Sheet web made of layers of parallel threads

PHOLCIDAE: *PHOLCUS*

CELLAR SPIDERS

Cellar spiders are common in almost any city, where they build their domed webs in ceilings, basements, or virtually any corner. In the wild they can be found under stones and overhangs, or in caves. They have extremely long legs, a small body, and a cylindrical abdomen. Because of their success in living in human constructions, they have been transported everywhere and have thus become cosmopolitan. It is not clear where they originated, but judging by their closest relatives, they may have originally inhabited the Middle East or central Europe.

GENUS
Pholcus

DISTRIBUTION
Cosmopolitan

HABITAT
In and around city buildings worldwide. In the wild, occurs under stones or in caves

CHARACTERISTICS
- Extremely long, thin legs
- Male palps robust, with tarsus modified into a complex structure

ABOVE LEFT | This female *Pholcus phalangioides* is carrying her eggs with the chelicerae. This species is a common inhabitant of houses, where it builds large sheet webs.

ABOVE | A male *Pholcus phalangioides*, with strong palps and complex copulatory organs. These spiders have eight eyes, two of them in the middle and the others in two triads at the sides.

Cellar spiders look fragile and are harmless to humans, but they are fearsome opponents for insects and other spiders. They swiftly attack their prey, wrapping it with a gluey, viscous silk, while maintaining a safe distance, thanks to their long legs. Cellar spiders are also aggressive mimics that invade other spiders' webs and imitate the vibration of a trapped insect; when the web owner comes after the supposed prey, it is soon entangled and subdued with viscous silk.

A peculiar behavior of cellar spiders is "whirling"; when disturbed, the spider swings so rapidly on its web that it becomes a confused blur to the human eye. Pholcid mating is also peculiar; the male locks the external female genitalia with his chelicerae and inserts both copulatory organs simultaneously. The female carries the egg sac with her chelicerae, and wraps it with relatively little silk, so that the spiderlings are visible while they carry out their first molts before hatching.

PLECTREURIDAE: *PLECTREURYS*

SPUR-LIPPED SPIDERS

Plectreurids are a small family of spiders containing two genera, *Plectreurys* and *Kibramoa*, restricted to dry habitats from the southern USA to Costa Rica and Cuba. The carapace and legs are heavily sclerotized. *Plectreurys* species are sedentary and build a horizontal web between rocks near the ground, with a hidden retreat. As with many desert spiders, *Plectreurys* can survive with no water source other than the insects they consume. Males have a clasping spur on the front tibiae, presumably used to immobilize the female during mating.

BELOW | A female *Kibramoa suprenans pima*. The species *K. suprenans* lives in California, USA. An isolated population in the Baboquivari Mountains in Arizona is currently distinguished as the subspecies *K. s. pima*, but is probably a separate species.

GENUS
Plectreurys

DISTRIBUTION
Southern USA to Costa Rica and Cuba

HABITAT
Dry, desert conditions; webs are built between rocks

CHARACTERISTICS

- Simple haplogyne female genitalia
- Eight eyes
- Males have a clasping spur on tibiae

DESERT CONE-WEB SPIDERS

Diguetia is one of the two genera of the small family Diguetidae. Nine *Diguetia* species live in deserts of the southwestern USA and Mexico, whereas a 10th species occurs far away in central Argentina—a biogeographic mystery. These spiders make a tangle web with a horizontal platform and a vertical silken cone in its center; the cone is used as a retreat and to house the egg sacs. The spiders stick debris to the silken retreat—the silk is extremely resistant and can endure very strong winds.

BELOW | A female *Diguetia* on its web, at Riverside County, California, USA. These spiders make very robust webs on desert shrubs. The occurrence of a single *Diguetia* species in the deserts of Argentina is a biogeographic mystery.

GENUS
Diguetia

DISTRIBUTION
Dry areas of USA, Mexico, and Argentina

HABITAT
Shrubs in dry deserts

CHARACTERISTICS
- Six eyes
- Haplogyne genital system
- Web has a horizontal platform and central conical retreat

CAPONIIDAE: *TARSONOPS*

BRIGHT LUNGLESS SPIDERS

Caponiids are among the weirdest of all spiders. They may have two, four, six, or eight eyes, but—unlike other soil-dwelling spiders with reduced eye numbers—they retain the anterior median pair of eyes. They lack the characteristic book lungs of spiders, instead having a thick bunch of tubular tracheae. Their legs have unique morphologies as well. Many caponiids have additional articulations, ridges of unknown function, and also the only soft adhesive pads found in any spider.

BELOW | A female *Caponina notabilis*, a colorful species from southern South America. Although caponiids are fairly diverse, they are among the rarest spiders, probably because of their cryptic habitats.

GENUS
Tarsonops

DISTRIBUTION
Southern USA to Panama, Cuba

HABITAT
Dark, hidden areas: leaf litter, soil, rock crevices, clumps of grass, beneath rocks and logs

CHARACTERISTICS

- Have only two dark eyes
- Posterior median spinnerets in an advanced position, situated between the anterior laterals
- Lack book lungs, and have thick bunches of tracheae instead

Caponiids are extremely rare in collections, and virtually nothing is known of their biology or diet. They are found in dark, cryptic habitats such as leaf litter, soil crevices, and grass tussocks, and under rocks and logs, where they spin a silken retreat. Many species have a red carapace and legs, and sometimes a blue abdomen. Caponiids are distant cousins of the dysderoid spiders (goblin spiders and relatives). The genus *Tarsonops* has seven species, all with only two eyes. They occur from California to Panama.

BELOW | A male *Tarsonops systematicus*, California Coast Ranges, USA. *Tarsonops* are unusual among spiders in that they have only two dark eyes.

TETRABLEMMIDAE: *TETRABLEMMA*

MINUTE ARMORED SPIDERS

Tetrablemma is a genus of minute armored spiders from the tropics of all continents, with most diversity in Southeast Asia. Like other tetrablemmids and also pacullids, the species have a heavily sclerotized body; the abdomen is covered by sclerotized plates, larger on the dorsum and venter, and smaller on the sides. These spiders can be very abundant in the leaf litter.

Although they come from a haplogyne lineage, the females have some of the most complex genital systems known in spiders. Little is known of their biology; they probably build horizontal sheet webs in the interstices of the leaf litter. Tetrablemmids are among the most bizarre-looking spiders; besides their armored body, some have only four eyes in the center of the carapace, horns on the carapace, or projections on the chelicerae.

RIGHT | A female *Tetrablemma sokense* from a cave in Battambang, Cambodia. Tetrablemmids are tiny spiders covered by hard, armor-like platelets. They are very difficult to detect but can be found by sifting through the leaf litter where they live.

GENUS
Tetrablemma

DISTRIBUTION
Pantropical

HABITAT
Leaf litter of tropical forests, in caves

CHARACTERISTICS
- Minute size
- Heavily sclerotized abdomen, covered by dorsal scutum and many lateral and ventral platelets
- Four eyes
- Complex female genital system

PACULLIDAE: *PERANIA*

GIANT ARMORED SPIDERS

Pacullids are heavily sclerotized medium-sized spiders that live in the rainforests of Southeast Asia. As in tetrablemmids, their abdomen is covered by hard platelets, and they were included in the same family until recently. *Perania* species are known from southern China, Myanmar, Thailand, Vietnam, Malaysia, and Indonesia. They build horizontal webs near the soil, most easily found on banks. The males of several *Perania* species have an anterior truncate horn projecting outward below the eye, whose function as yet remains unknown.

LEFT | A male giant armored spider of the genus *Perania*, from the Cameron Highlands, Malaysia. Pacullids have the strongest armor of all spiders, and the males may have bizarre projections in front of their eyes.

GENUS
Perania

DISTRIBUTION
Southern China to southern Sumatra

HABITAT
Rainforests, in the soil and among leaf litter

CHARACTERISTICS

- Heavily sclerotized abdomen, covered by dorsal scutum and many lateral and ventral platelets
- Six eyes
- Simple haplogyne genital system

TROGLORAPTORIDAE: *TROGLORAPTOR*

CAVE ROBBER SPIDER

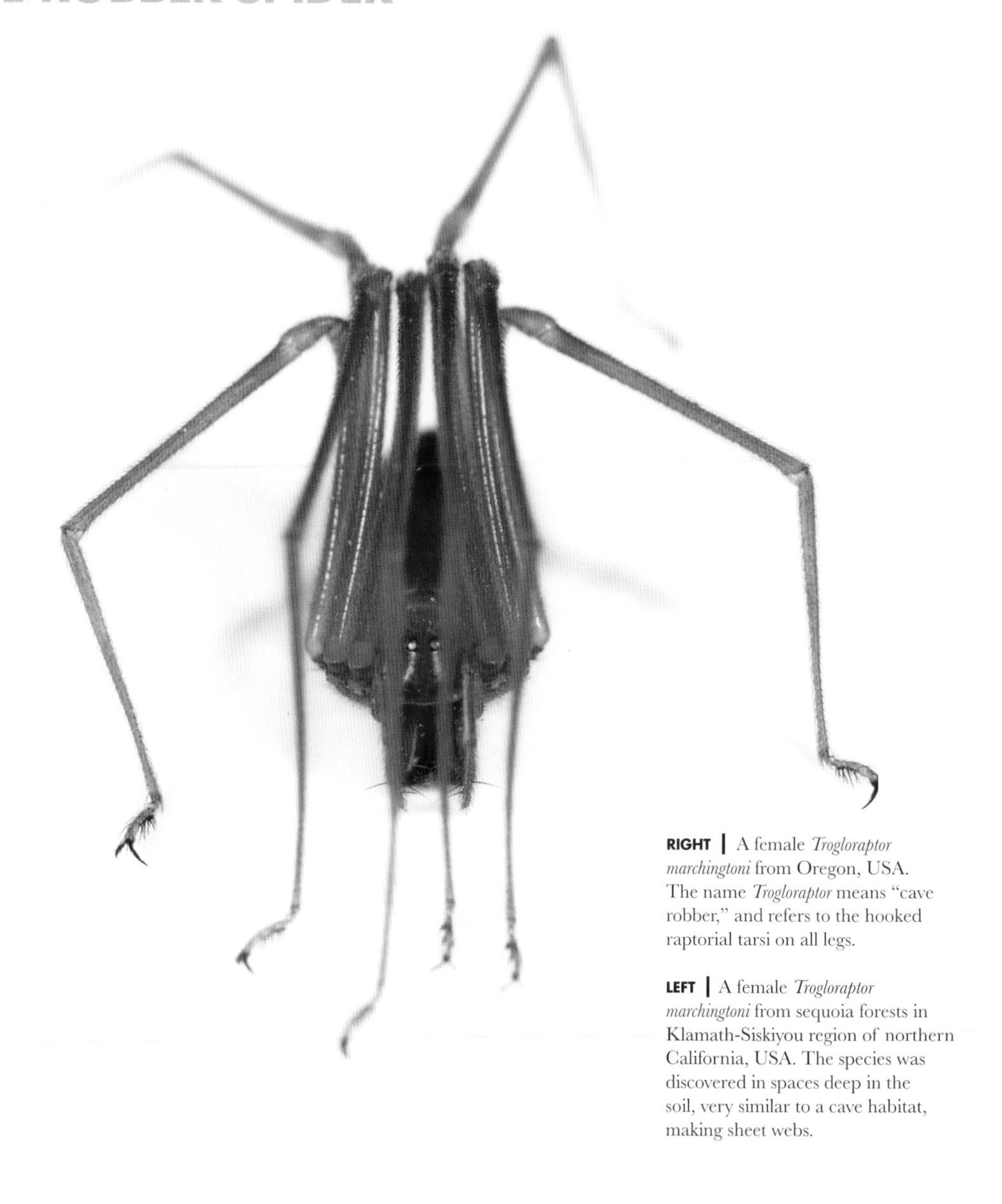

RIGHT | A female *Trogloraptor marchingtoni* from Oregon, USA. The name *Trogloraptor* means "cave robber," and refers to the hooked raptorial tarsi on all legs.

LEFT | A female *Trogloraptor marchingtoni* from sequoia forests in Klamath-Siskiyou region of northern California, USA. The species was discovered in spaces deep in the soil, very similar to a cave habitat, making sheet webs.

GENUS
Trogloraptor

DISTRIBUTION
Southwestern Oregon and northern California, USA

HABITAT
Caves, or deep in soil spaces in sequoia forests

CHARACTERISTICS
- Asymmetric, raptorial claws and flexible tarsi on all legs
- Six eyes
- Simple male genitalia

Trogloraptor marchingtoni is a mysterious spider that was recently discovered in caves in southwestern Oregon and deep in soil spaces in old-growth sequoia forest in northern California. The species seems to be the sole surviving member of an ancient lineage of spiders, and thus belongs in its own separate family. *Trogloraptor* specimens hang from horizontal webs on the roof of caves or in secluded spaces deep in the forest soil. They have large claws on all their legs, and additional articulations on the tarsi, which allow the claws to be folded against strong spines, forming raptorial appendages.

The spinneret morphology of *Trogloraptor* suggests they are relatives of dysderoids (the group including dysderids, segestriids, orsolobids, and oonopids). Overall, however, they have a remarkably primitive morphology. The mouthparts are also odd. In most araneomorphs, the basal segment of the pedipalps (the endites), which face the posterior side of the chelicerae, bear a serrula near their tip that is composed of a single row of teeth; in *Trogloraptor*, the serrula consists of two rows of teeth. As with many interesting taxa in the Pacific Northwest, *Trogloraptor* is a relict species—the unique, primitive relative of a widespread and successful group of spiders.

SEGESTRIIDAE: *SEGESTRIA*

TUBE-WEB SPIDERS

Segestria spiders belong to a rather small family (Segestriidae) with four genera and 130 species worldwide. The genus itself contains almost 20 species, found mainly in the Holarctic (North America and Eurasia). All members have six eyes arranged in three pairs. Another typical character is the leg position: six legs are directed to the front, and only the fourth pair is directed backward. This unusual arrangement reflects the spiders' lifestyle: they inhabit tubular nests with two openings. At the front, to the outside of the environment, radial threads are built as fishing-wire traps for running arthropods, conducting vibrations back to the tube and its occupant. The nocturnal spiders ambush prey at the entrance of their burrow, holding six legs in close contact with the signal lines. When disturbed, they hide in the burrow and if necessary escape through the hind exit on the other side.

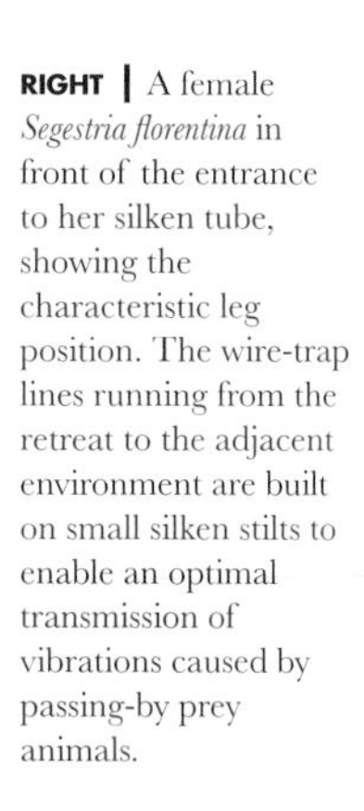

RIGHT | A female *Segestria florentina* in front of the entrance to her silken tube, showing the characteristic leg position. The wire-trap lines running from the retreat to the adjacent environment are built on small silken stilts to enable an optimal transmission of vibrations caused by passing-by prey animals.

GENUS
Segestria

DISTRIBUTION
Primarily Holarctic; one species introduced to South America (Brazil, Uruguay, Argentina)

HABITAT
Tubular nests behind tree bark, between stones and similar places; in natural habitats and on human dwellings

CHARACTERISTICS
- Six eyes arranged as three pairs
- Anterior three pairs of legs directed to the front, with only the fourth pair directed backward
- Construct open tubes as retreats

Ariadna species have a similar lifestyle to that of *Segestria*, but one exhibits a unique behavior. This species, found in the Namib Desert, builds vertical tubes in the sand and places four to ten small stones around the tube entrance. In most cases, the entrance is ornamented with seven stones, hence the common name of the species, Seven-Stone Spider. The function of the stones is not yet known.

ABOVE | A female *Segestria florentina* showing the characteristic eye arrangement of the family and the specific green iridescent coloration of the cheliceraeof this particular species. These spiders will defend themselves and are known to bite human fingers when they feel threatened.

DYSDERIDAE: *DYSDERA*

SIX-EYED SPIDERS

GENUS

Dysdera

DISTRIBUTION

Palearctic (note: the map includes *D. crocata* C. L. Koch, 1838, indigenous to Europe, Caucasus, Iraq, and central Asia); introduced to North America, Chile, Brazil, Australia, New Zealand, Hawaii

HABITAT

Under stones and logs in a variety of habitats, e.g., open land, forest, caves

CHARACTERISTICS

- Six eyes grouped together in two rows, anterior row with two eyes only
- Chelicerae often specialized for capturing woodlice (isopods)
- Reddish coloration

Dysdera spiders are small to medium-sized nocturnal ground-dwellers that hide during the day under stones and logs. There are more than 280 described species, found in Europe, northern Africa, and Asia. They are characterized by six eyes and the chelicerae, which protrude to the front. With their reddish appearance, these spiders sometimes elicit fear in humans, but their venom is not harmful.

Some *Dysdera* species specialize in eating woodlice—including even the pillbugs that try to deter predators by rolling up to form a ball, with only their tough dorsal sclerites showing. Researchers have discovered that the shape of the spider's fangs correlates with different strategies to overcome the hard exoskeleton of these crustaceans. There are flattened fangs that can be inserted between the sclerites; elongated fangs, where one fang can reach to the softer underside of the woodlouse while the other holds the prey's dorsal side; and even dorsally concave fangs, both used for biting the underside of the woodlouse.

When a male and female meet, the male takes the initiative and grasps the female's pedicel with his fangs. He touches the female's abdomen with his forelegs, and eventually inserts his palps to transfer sperm into the female's spermathecae. In a related species, *Harpactea sadistica*, the male pierces the female's abdomen at the ventral side to insert the sperm directly into the body cavity. Usually, spider eggs are fertilized outside of the body as they are laid. In *H. sadistica*, however, the eggs are fertilized in the ovary and laid as embryos. Such a traumatic insemination may help the male secure his paternity, since females of other species sometimes have strategies to remove sperm stored in their spermathecae in favor of a more recent mate.

LEFT | Close-up of the front of a male *Dysdera crocata*, showing the typical eye arrangement in which the anterior row has only two eyes and the procurved posterior row includes four eyes.

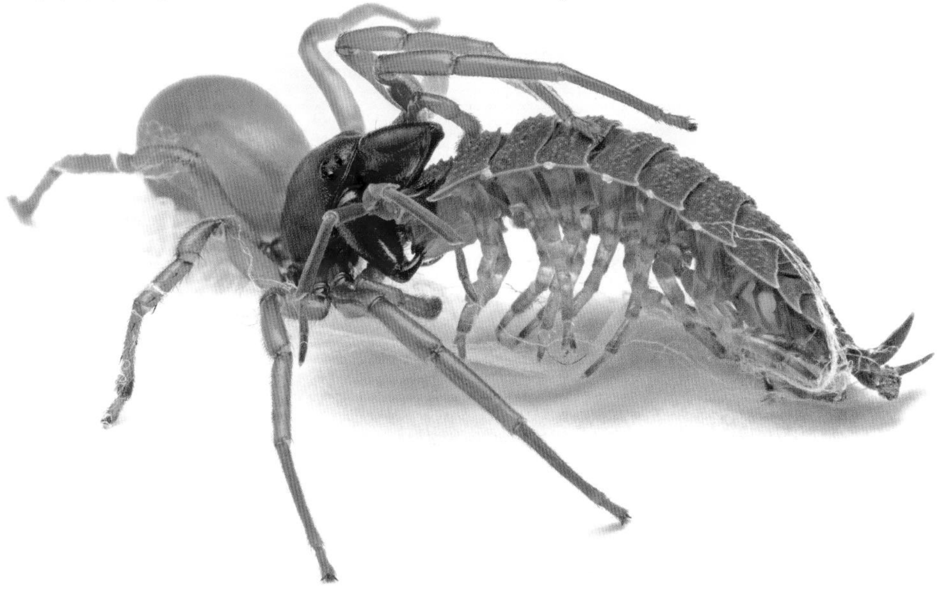

BELOW | A female *Dysdera crocata* feeding on a Rough Woodlouse (*Porcellio scaber*). The right fang pierces the underside of the otherwise heavily armored woodlouse and paralyzes the prey by injecting venom.

OONOPIDAE: *TRIAERIS*

GOBLIN SPIDERS

Oonopids are one of the largest groups of spiders, currently including 114 genera and more than 1,750 species. Our knowledge of these tiny, usually nocturnal hunting spiders has advanced greatly in recent years. Goblin spider species have what may be the smallest average size range of any spiders. For example, in the genus *Costarina*, which occurs from Mexico south to Colombia, the small nation of Costa Rica is home to more than 50 species—all but three occur nowhere else and are restricted to very small parts of the country. However, about a dozen goblin spider species provide notable exceptions to this pattern. These are typically synanthropic (living around human settlements) and can attain pantropical distributions, including many far-flung islands.

Triaeris stenaspis is one of these examples; although the genus appears to be originally restricted to West Africa, this species is very widely distributed. It is one of the very few spiders known to be parthenogenetic; in the lab, females that have never mated can lay viable eggs, and entire generations have been raised in the total absence of males. Many thousands of

RIGHT | *Triaersis stenaspis* is minuscule, and only females have ever been found. The species has been described as "an enigma wrapped in a mystery," and is very unusually parthenogenetic, reproducing in the absence of males.

GENUS
Triaeris

DISTRIBUTION
Probably originally from West Africa; spread by humans and now pantropical

HABITAT
Forest litter; the synanthopic species even occurs in European greenhouses

CHARACTERISTICS
- Abdomen topped with a hard plate
- One parthenogenetic species with no records of any males
- Six eyes, clustered closely together

specimens have been collected, but no males of the species have ever been found. Because only the transport of a single female is required to found a new population, the species now occurs in such remote places as the Marquesas and Cook Islands; populations have even become established in European greenhouses.

ABOVE | In virtually all male spiders, the right and left sperm-transferring palps used in mating are exact mirror images of each other, but in this species, *Paradysderina carrizal* from Colombia, and a few other goblin spiders, the right and left palps are different (and consistently so, within a species).

OONOPIDAE: *ORCHESTINA*

JUMPING GOBLIN SPIDERS

O*rchestina* is a large and almost worldwide genus of goblin spiders that currently contains more than 160 species. Although most oonopids are ground-dwelling, hiding and hunting their prey in leaf litter, species of *Orchestina* have conquered a very different habitat, forming one of the most abundant parts of the spider fauna in tropical forest canopies. Sampling the arthropods living high above ground at the tops of trees is relatively new, enabled by "canopy fogging." In this process large sheets are laid out at ground level, and a machine is used to produce a cloud of steam that includes pyrethrum, an insecticide. Animals that come into contact with the spray fall onto the sheets below, where they can be collected, without the researchers having to climb dozens of feet to the tops of the trees.

Perhaps the most distinctive feature of *Orchestina* is the enlarged femur found on their

ABOVE | A male *Orchestina* from Australia; note the enlarged femora on the fourth pair of legs, which contain sizable muscles, allowing the animal to jump several times its body length.

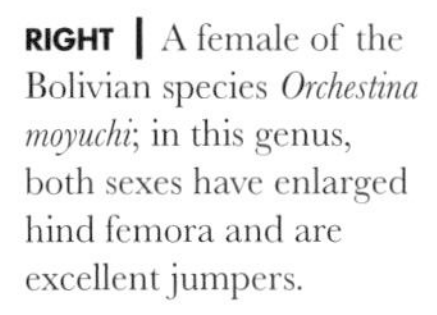

RIGHT | A female of the Bolivian species *Orchestina moyuchi*; in this genus, both sexes have enlarged hind femora and are excellent jumpers.

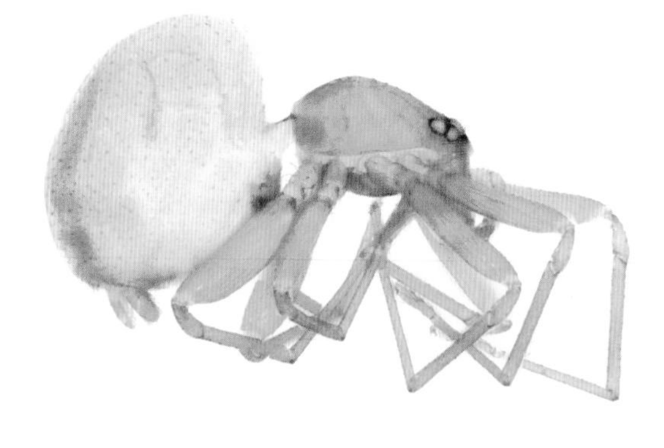

GENUS
Orchestina

DISTRIBUTION
Northern USA south to the northern half of Chile and Argentina; all Africa; Tasmania; southern Europe to Japan

HABITAT
Forests, within leaf litter; some species occur in tropical forest canopies

CHARACTERISTICS
- Enlarged femur on posterior legs
- Six eyes, in a transverse row
- Legs lack spines

LEFT | A male of the North American species *Escaphiella hespera*; members of this genus and the closely related *Scaphiella* are commonly called taco spiders because of the unique conformation of their abdominal plates. In both sexes, the plate on the underside of the abdomen extends completely up both its sides, forming a structure shaped like a taco shell. In adult males, another plate covers the top of the abdomen, making them rather crunchy tacos.

BELOW | A scanning electron microscope image showing a dorsal view of a female *Escaphiella hespera*. Note the soft cuticle separating the right and left sides of the extended plate on the abdomen (i.e., the winglike structures).

fourth pair of legs. Using the muscles in these sturdy leg segments, the spiders are able to jump several times their body length. When a spider happens across your desk, it takes only a slight disturbance, especially from the rear, to induce the animal to jump; it can then repeat the process quickly, enabling it to escape from some potential predators.

The genus appears to be one of the most primitive groups of goblin spiders. Most members of the family have lost the distinct sperm duct that is found inside the palps of males of all other spiders. However, males of *Orchestina*—and a few other goblin spider genera—still retain that feature. Their eye pattern, with the six eyes arrayed transversely, in an H-shape, is also primitive (in most oonopids, the lateral pairs of eyes are much closer together). Interestingly, *Orchestina* species are also by far the goblin spiders most commonly found as fossils in amber.

Although *Orchestina* and many other basal goblin spiders are soft-bodied, other goblin spiders are notably harder, with strong plates on the upper and/or lower surfaces of the abdomen. Indeed, the family was classically divided into two subfamilies on the basis of whether or not the abdomen has such plates. That classification proved to be artificial, however; there are even goblin spider species where the males have abdominal plates but the females do not.

ORSOLOBIDAE: *CORNIFALX*

ORSOLOBID GOBLIN SPIDERS

BELOW | This attractively patterned species belongs to *Cornifalx*, known only from Tasmania; males have bizarre horns at the front of their carapace and chelicerae.

Orsolobids are a classic Gondwanan group. Their modern distribution is restricted to the southern continents that were once united as Gondwana, namely Australia, Tasmania, southern South America (Chile, Argentina, southern

GENUS
Cornifalx

DISTRIBUTION
Tasmania; other genera occur elsewhere in the southern hemisphere

HABITAT
Forests, often within leaf litter; sometimes underground

CHARACTERISTICS
- No more than six eyes
- Pair of posterior spiracles situated just behind epigastric furrow
- Legs with elevated tarsal organs

Brazil), the Falkland Islands, and southern Africa (Malawi, South Africa). Some 30 genera and more than 175 species are recognized; members of the genus *Cornifalx*, shown here, are known only from Tasmania.

BELOW | A scanning electron micrograph image of an anterior view of a male *Cornifalx insignis* from Tasmania, showing the large median "horn" on the carapace, as well as two smaller projections on the sides of the carapace and two others on the chelicerae.

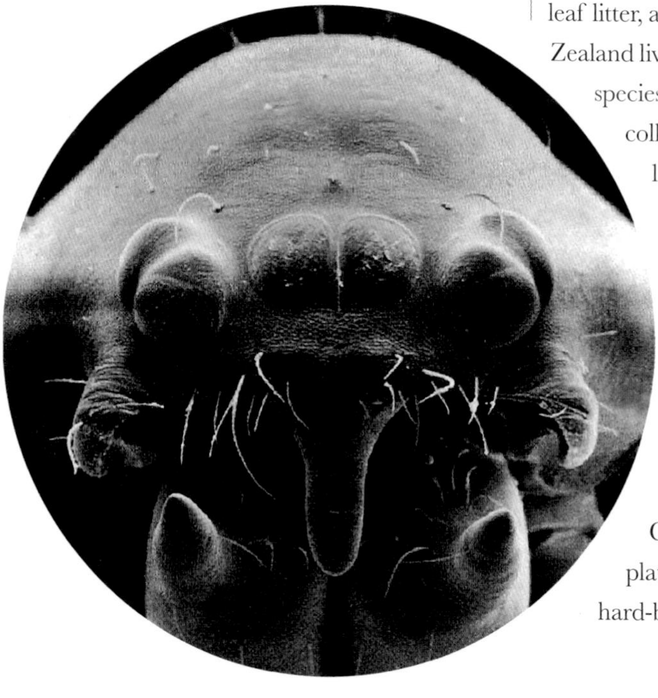

As in oonopids, dysderids, and segestriids, the posterior spiracles of orsolobids, which lead to the respiratory tracheae, are situated far forward on the abdomen, just behind the spiracles of the book lungs, and the female genitalia have both anterior and posterior elements. What sets orsolobids apart from the other families—and most other spiders—is the unusual structure of their tarsal organs. These tiny structures, situated near the tips of the leg and palpal tarsi, are chemosensors that are usually flat. However, in orsolobids they are elevated above the surface of the legs, often enough to make them observable with a light microscope (usually, a scanning electron microscope is needed to see these minute features).

Orsolobids are vagrant hunters, although at least some species construct a flimsy silken retreat, with females placing their flattened egg sacs near the back of the retreat. They are often found in leaf litter, and at least one species from New Zealand lives far below the ground surface. That species lacks eyes entirely, and has only been collected once, in a trap set more than 13 feet (4 m) below the surface; the trap was intended to collect aquatic organisms, but an unexpected dry spell allowed it to sample the fauna of tiny organisms that live interstitially in the ground, including this unpigmented, blind species.

Two of the orsolobid genera, *Duripelta* from New Zealand and *Basibulbus* from Chile, have male abdomens covered by plates, and thus can easily be confused with hard-bodied oonopids.

PALPIMANIDAE: *PALPIMANUS*

PALP-FOOTED SPIDERS

Palpimanids are a group of 18 genera and almost 150 species, found from Cuba to Argentina, throughout Africa, and across the southern parts of Eurasia. They can usually be recognized easily by their highly modified front legs. The femora of that pair of legs are dorsally enlarged, massive, and sometimes even provided with spurs; the distal segments (the tibiae, metatarsi, and tarsi) are each equipped with a peculiar pad of specialized, adhesive hairs. Each of those hairs has an expanded tip bearing dozens of tiny projections. These leg modifications apparently help palpimanids safely capture their favorite prey—other spiders. Some species appear to specialize on capturing jumping spiders, whose vision is vastly better; they will even enter their prey's silken retreats. When they encounter the retreat of a jumping spider, they will pull and chew on the silk, making and enlarging a hole. Once in the retreat, they will attack and feed on the jumping spider, whose motion is somewhat limited by the walls of the retreat. In some cases, they will also feed on the eggs of the salticid.

RIGHT | Side view of a palpimanid from Mozambique; note the massive femur and long patella on the greatly enlarged front pair of legs.

GENUS
Palpimanus

DISTRIBUTION
Africa, Mediterranean to central Asia and India

HABITAT
Leaf litter in dry and humid forests

CHARACTERISTICS
- Leg I typically enlarged, especially the femur
- Distal segments of leg I with prolateral pads of specialized hairs
- Reduced spinnerets, usually numbering only two in females

ABOVE | A *Palpimanus* from Mozambique, displaying the enlarged front legs; note the pads of adhesive hairs on the distal segments of those legs.

Other peculiarities of the group include reduced spinnerets. Often there is only one pair of spinnerets (unlike typical spiders, which have three pairs). If the posterior median and lateral spinnerets are represented at all, it is just by spigots originating directly from the surface of the abdomen, rather than from distinct spinnerets.

Of the four genera found in the New World, perhaps the most unusual is *Fernandezina*, which occurs widely in South America. In that genus, the first femur is not nearly as expanded as in other palpimanids, and the animals are sexually very dimorphic. The males have an elongated plate on the top of the abdomen that is absent in females.

STENOCHILIDAE: *COLOPEA*

DIAMOND-HEADED SPIDERS

Stenochilids seem to be close relatives of palpimanids, sharing with them the pads of specialized hairs on the front legs (although in stenochilids, the pads also occur on the second pair of legs, and are divided into lateral and ventral portions). It is not known whether stenochilids also prefer other spiders as their prey, but the pads could certainly help them hang onto fast-moving prey.

Stenochilids are easily recognized by their peculiar carapace. This is distinctly diamond-shaped (i.e., much wider in the middle than at either the front or back), and has two thoracic grooves—a slit-shaped anterior one, like that found in most other spiders, plus a more posteriorly situated pit. Like palpimanids, there are two large spinnerets. Unlike them, however, there are still remnants of the two posterior pairs of spinnerets; these remnants seem to be functional, at least in females.

The family contains only two genera: *Stenochilus*, known only from India, Sri Lanka, Myanmar, and Cambodia; and *Colopea*, which is more widely distributed, from China through Southeast Asia to Borneo, the Philippines, Bali, Fiji, New Guinea, and northern Australia.

In *Stenochilus*, the carapace is even more oddly shaped, with lateral undulations.

LEFT | A juvenile *Colopea* from northern Australia; note the unusual diamond-shaped carapace.

GENUS
Colopea

DISTRIBUTION
China, south to northern Australia

HABITAT
Leaf litter in dry and humid forests

CHARACTERISTICS
- Distal segments of legs I and II with prolateral pads of specialized hairs
- Diamond-shaped carapace with two thoracic grooves
- Six spinnerets, but posterior two pairs reduced in size

NEW ZEALAND PALP-FOOTED SPIDERS

Huttoniids are one of the few spider families whose known distribution is restricted to a single country—in this case New Zealand. Only one species has been described, and few specimens were known until their favored habitat was discovered; they prefer to live in the dead fronds of low-growing, rainforest ferns.

The newly available collections from those habitats show that there are around 20 additional species, and possibly a second genus. They construct a purse-like retreat in the fern fronds, leaving it to hunt for their prey. At least in captivity, huttoniids will feed on a wide variety of insects, as well as on the terrestrial amphipods common in the moist, decaying fronds where they live. Their chelicerae lack the marginal teeth found in most spiders, but have numerous peg teeth that are socketed rather than originating directly from the cuticle. In contrast, palpimanids usually have both normal and peg teeth, whereas stenochilids have neither.

LEFT | A male *Huttonia* from New Zealand, showing the chevroned color pattern on the abdomen.

GENUS
Huttonia

DISTRIBUTION
New Zealand

HABITAT
Rainforest, within dead fronds of low-growing ferns

CHARACTERISTICS
- Distal segments of leg I with a row, rather than pad, of specialized hairs
- Male palp with a dense, prolateral pad of specialized setae
- Six spinnerets

MECYSMAUCHENIIDAE: *MECYSMAUCHENIUS*

TRAP-JAW SPIDERS

Mecysmaucheniids are among the most unusual-looking spiders: their carapace is greatly elevated, and extends completely around the elongated chelicerae. The group is found only in southern South America (Chile and Argentina, including the Juan Fernández and Falkland Islands, with five genera) and New Zealand (with two genera).

In their normal position, the chelicerae extend down and are held close to the body, but when the spider is hunting, they are widely spread, extending far to the sides of the carapace. A row of tiny hairs extends down the inner side of the chelicerae; when prey contacts these hairs, the chelicerae snap together, with astonishing speed. High-speed video shows that in the slower species, this happens in

RIGHT | A male *Mecysmauchenius fernandezi*, with the chelicerae in hunting position.

GENUS
Mecysmauchenius

DISTRIBUTION
Chile and Juan Fernández Islands, Argentina and Falkland Islands, other genera in New Zealand

HABITAT
Southern beech forests, within damp moss; humid grasslands, in moist tussocks

CHARACTERISTICS
- Carapace greatly elevated, forming a circular opening around the chelicerae
- Chelicerae greatly elongated, opening widely to the sides
- Only two spinnerets

ABOVE | A female *Mecysmauchenius* with her egg sac; the elongated chelicerae are visible at the front.

about 0.02 seconds; in the fastest species, it is two orders of magnitude faster, taking a mere 0.00012 seconds. This is by far the fastest predatory strike recorded for spiders. The extraordinary speed is "power-amplified," meaning that the force exerted far surpasses ordinary muscle movement; instead, energy that is stored slowly in the muscles is released almost instantaneously. Power amplification occurs in some other arthropods, especially ants, but these are the only spiders in which it has been demonstrated.

Mecysmaucheniids are usually found in southern beech forests, living in damp moss; they also occur in grasslands where grass tussocks retain sufficient humidity. Although they will eat small spiders, at least in captivity mecysmaucheniids also take a wide variety of insects and other invertebrates. At least some species spin silk retreats in the moss and place their egg sacs in those retreats. Two subfamilies are recognized, differing in their cheliceral dentition. Interestingly, they are not geographically distinct: there is a close relationship between Chilean and New Zealand genera within each subfamily. The only species that still retains all eight eyes is from New Zealand, whereas the others possess just six eyes.

ENTELEGYNAE

Entelegynae is a large subgroup of true spiders, including more than 70 families and about 34,000 species (out of more than 115 families and about 48,000 species).

Most spider species are distinguished by taxonomists through their differently developed genitalia. But these complex structures also offer clues about their higher-level relationships; morphological innovations within the reproduction system can define entire groups of spiders. The Entelegynae (from the Greek for "females with complete genitalia") comprise a large group of families that share a special arrangement of the female copulatory organ.

LEFT | Bold Jumping Spider (Salticidae, *Phidippus audax*), common in North America.

SPIDER FERTILIZATION SYSTEMS

The reproductive tracts of female spiders are either haplogyne or entelegyne. The more ancestral groups of araneomorphs possess simple ("haplogyne" or "non-entelegyne") genitalia, in which the eggs are fertilized with sperm that travel both to and from the spermathecae through a single genital opening. Entelegyne genitalia are more advanced, as they possess an extra pair of copulatory ducts leading from the outside to the spermathecae.

HAPLOGYNE OR NON-ENTELEGYNE

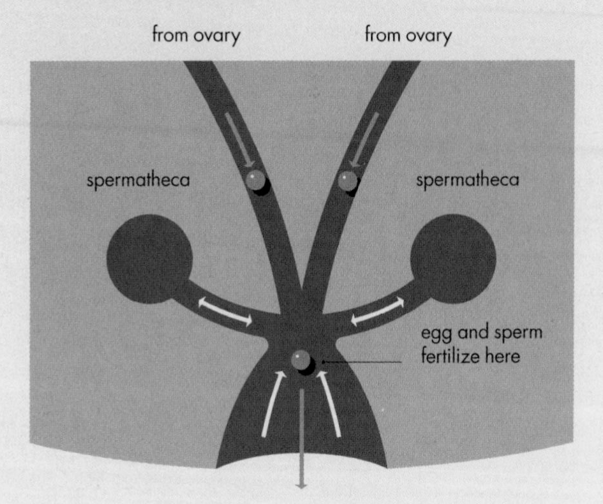

Single genital opening

ENTELEGYNE

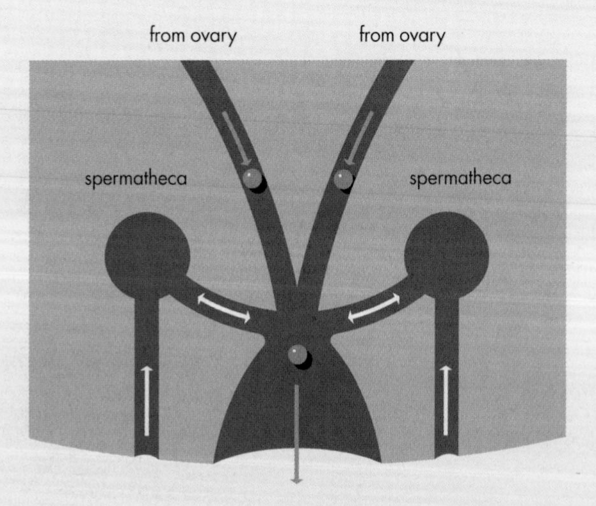

An extra pair of copulatory ducts lead directly to the spermathecae

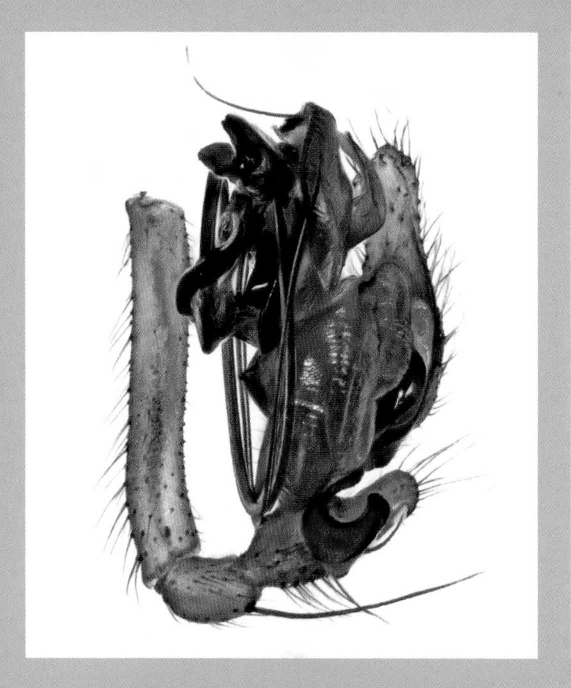

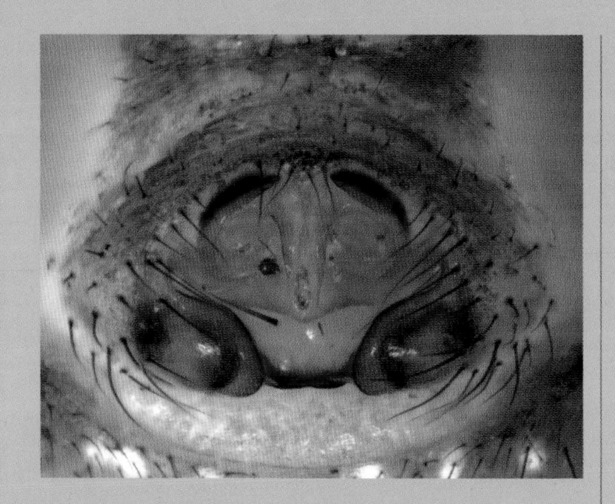

ABOVE | The male and female genitalia of *Labulla thoracica*, a European linyphiid species, showing the complexity of these structures. Above, lateral and ventral views of the left male palp; below, ventral view of the female epigynum.

Two additional pores are found close to the primary genital opening (i.e, the slit through which the eggs are laid). These copulatory openings are situated on a sclerotized plate (the epigynum) and connect with the internal spermathecae, in which the sperm of the male is stored until the egg sac is built. Through another pair of ducts (called the fertilization ducts), sperm move into the uterus externus; here they fertilize the eggs while they are being deposited in the egg sac. In males, accordingly complex copulatory organs with various sclerites and an amazing diversity of structures, conformations, and mechanics co-evolved as counterparts to the female's epigynum.

MEGADICTYNIDAE: *MEGADICTYNA*

LARGE SPINNING FIELD SPIDERS

The single known species of *Megadictyna*, *M. thilenii*, can be found in the wet forests of both the North and South Islands of New Zealand. They are found under logs and stones, where they build a small regular sheet web with cribellate silk; the spider hangs underneath the web.

Whether *Megadictyna* actively use their small webs for foraging is unknown, but at night they can sometimes be found wandering on the forest floor. The only other member of this family is *Forstertyna marplesi*, known from just a few female specimens collected on the South Island of New Zealand.

BELOW | A male *Megadictyna thilenii*, from the South Island of New Zealand. The female is similar in appearance, but slightly larger in size.

GENUS
Megadictyna

DISTRIBUTION
New Zealand

HABITAT
Floor of damp forests, under logs and stones

CHARACTERISTICS
- Medium to large entelegyne spiders with an entire cribellum and an enlarged spinning field on the posterior lateral spinnerets
- Male palpal tibia with large, curved dorsal apophysis
- Respiratory system with a pair of book lungs, a pair of branched median tracheae, and a pair of simple lateral tracheae

RED-AND-BLACK SPIDERS

Nicodamids include 27 species in seven genera; most are known only from Australia, but five species have been described from New Guinea. Their conspicuous red (or bright orange) and black coloration makes nicodamids very distinctive. These spiders can be found in the leaf litter as well as under logs and stones. They build weak, irregular sheet webs at ground level.

BELOW | A male *Oncodamus* from New South Wales, Australia, displaying the characteristic black-and-red coloration of nicodamids.

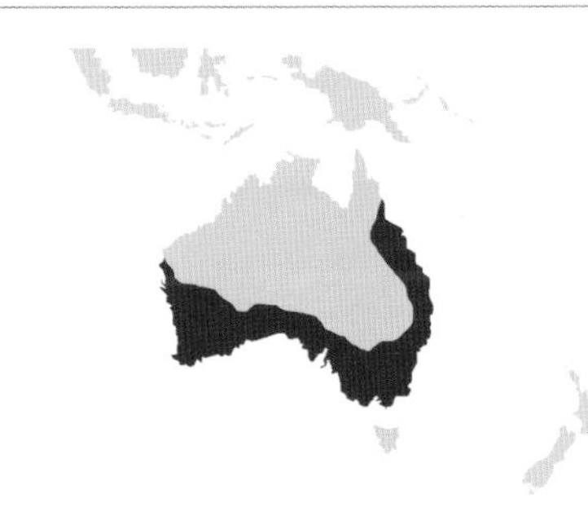

GENUS
Nicodamus

DISTRIBUTION
Australia; other genera in New Guinea

HABITAT
In leaf litter, mainly of forests; under logs and stones

CHARACTERISTICS
- Small to medium-sized entelegyne ecribellate spiders
- Male palpal tibia with dorsal and additional median apophysis
- A row of three or four stiff, dark setae in an otherwise large, bare area on the dorsal surface of the anterior lateral spinnerets
- Bright red carapace, legs, and sternum; black/dark purple abdomen

ERESIDAE: *STEGODYPHUS*

SOCIAL VELVET SPIDERS

Although some of its species are solitary, most *Stegodyphus* show striking social behavior. The average *Stegodyphus* nest is somewhat smaller than a football, but may contain up to several hundred individuals. The nest itself is made of very tough cribellate silk; certain African birds of prey incorporate these complex webs in their nests to provide stability. Thanks to their cooperative behavior, these spiders are able to overpower large prey—often many times larger than the spiders themselves. The females do all the work; apart from propagation, males do not collaborate in any of the colony's activities. The females are dull in color and well camouflaged, whereas males often have a spectacular color pattern.

In general, spiders tend to be particularly aggressive, even toward conspecifics (members of the same species). The existence of sociality and collaboration among spiders has therefore elicited much interest among scientists, and the behavior of *Stegodyphus* has been studied in great detail. It has been shown that individuals in a nest divide the tasks according to their personalities, especially their degree of boldness and aggression. These

RIGHT | Males of the African Social Velvet Spider (*Stegodyphus africanus*) have a strongly contrasted pattern: it explains their absence from activities outside the nest where the camouflaged females do all the work.

GENUS
Stegodyphus

DISTRIBUTION
Africa through the Middle East to China; one locality in Brazil

HABITAT
Semi-arid shrubland, grassland

CHARACTERISTICS
- Several social species that construct communal cribellate webs
- Remarkable sexual dimorphism, with dull-colored females and spectacular males
- Rectangular carapace

ABOVE | Nests of social velvet spiders (*Stegodyphus*) are made of tough silk and incorporate some dry leaves to make them less conspicuous; most of the maintenance work is carried out by females by night, but large prey may be captured during the day.

characteristics not only play an important role in the success and survival of the colony, but also in the size of the adult females—some of which can be twice the size of others.

In regions of South Africa, nests of *Stegodyphus mimosarum* were brought inside and hung in kitchens or other fly-infested rooms to reduce the density of the insects. In some dairy farms, up to 80 nests were imported, reducing the fly population to a tiny fraction of their initial numbers and helping to increase the milk production of the cows.

ERESIDAE: *ERESUS*

LADYBIRD SPIDERS

In contrast to their social relatives, *Eresus* species are strictly solitary. The few of the 20-odd species that have been studied prefer sun-exposed habitats with low vegetation and light, well-drained soils where they can easily construct their tubular retreat. This silk-lined tube is protected and concealed by a dark, soil-colored silken mat. The females may live up to four years and spend their entire life underground. Their webs serve as signal systems mainly designed to reveal relatively large prey, which are overpowered by the spider's short but strong legs. They apparently mate only once and feed the few spiderlings—usually not more than 20—until they die, whereupon they are consumed by their offspring. Adult males roam around by day and are particularly brightly colored, a phenomenon known as aposematism. Aposematic colors are usually eye-catching, meant to warn potential predators about the poor palatability of the target. *Eresus* males vaguely resemble ladybird beetles, which are renowned for their foul taste. Although the taste of the *Eresus* themselves is probably acceptable to most predators, their resemblance to ladybird beetles is enough to protect them during their search for a partner.

Few spiders deserve the epithet "cuddly," but if there is an exception, it is doubtlessly the male *Eresus*. In several countries within the species' range, they are among the rare spiders that have received protected status. They earned it, not only because they are usually rare and their habitat is uncommon, but also because the males astonish naturalists with their exceptional color pattern.

LEFT | The activity period of an adult male ladybird spider (*Eresus sandaliatus*) is very short, lasting only a few weeks in a year. Their diurnal activity is highly dependent on weather conditions: they prefer warm, sunlit days.

GENUS
Eresus

DISTRIBUTION
Mainly the western Palearctic, including western Siberia and northern China

HABITAT
Dry areas with short vegetation and well-drained soil

CHARACTERISTICS
- Sturdy spiders
- Solitary species with a tubular retreat covered by a silken mat
- Remarkable sexual dimorphism, with dull females and spectacular males

FLATTENED ANT SPIDERS

Penestomus includes nine described species endemic to southern Africa and is the sole genus in the family Penestomidae. *Penestomus* are small, cribellate spiders with a flattened body; they are very rarely collected and their biology is very poorly known. Until recently, their genealogical affinities were thought to be near the velvet spiders (Eresidae). Following the discovery of male specimens and molecular data, however, it seems that *Penestomus* are more closely related to the zodariids (ant spiders).

Penestomus montanus, known from the Eastern Cape Province of South Africa and Lesotho, builds webs underneath rocks lying on the soil surface. The webs follow a winding path and reach about 7 inches (18 cm) in length. Prey remnants found in the webs suggest that *Penestomus* feed primarily on ants, along with small beetles, bugs, and crickets.

ABOVE | A female *Penestomus egazini* from South Africa; the flattened body helps these spiders live in narrow spaces under stones and bark.

RIGHT | A female of the same species, taken with its egg sac under the bark of an introduced *Eucalyptus* tree.

GENUS
Penestomus

DISTRIBUTION
South Africa, Lesotho

HABITAT
Beneath rocks overlying the soil surface

CHARACTERISTICS
- Cribellate entelegyne spiders with eight eyes
- Flat body, 0.12–0.24 inches (3–6 mm) long, with subrectangular carapace and clypeal hood (an extension of the clypeal margin over the chelicerae)
- Male palp with retrolateral tibial apophysis
- Posterior lateral eyes less than three eye diameters behind the posterior median eyes

OECOBIIDAE: *OECOBIUS*

ROUND-HEADED SPIDERS

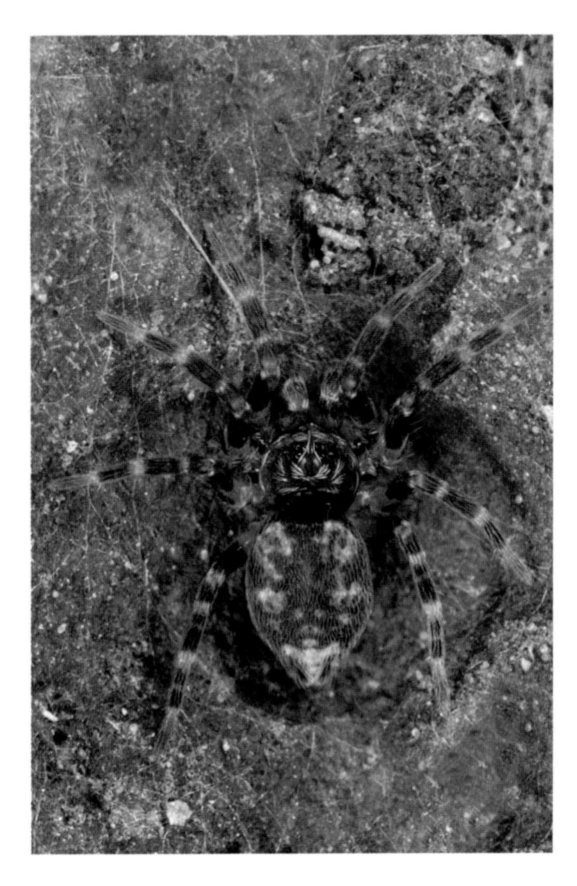

ABOVE | An *Oecobius* from Costa Rica, displaying its characteristic, almost circular carapace.

The family Oecobiidae includes about a hundred described species classified in six genera. Oecobiids are found on all continents except Antarctica. Some, such as *Oecobius navus* and *O. cellariorum*, are synanthropic and can be found in buildings around the world. *Oecobius* has evolved an insular radiation of more than 40 species in the Canary Islands. These small spiders build flat, densely meshed, star-shaped platform webs, consisting of two horizontal sheets, on walls and in crevices, with radiating trip lines. The spider sits under the web, between the two sheets, and is visible through the translucent silk film.

Oecobiids have a very characteristic carapace shape—wider than long, almost circular, with a projecting anterior margin, and lacking a fovea. The family includes cribellate species (such as those in the genus *Oecobius*) and ecribellate species (such as the large and beautifully colored Mediterranean species *Uroctea durandi*).

Oecobiids also have a uniquely modified anal tubercle with a double fringe of stout, curved setae, which is used in their specialized hunting technique. Once a prey (often an ant) is detected, the spider comes out from under the web and begins to circle around the prey. It applies a swathe of silk from the posterior spinnerets, which it combs using the setae

GENUS
Oecobius

DISTRIBUTION
Cosmopolitan

HABITAT
On walls and rocks, sometimes in crevices

CHARACTERISTICS
- Carapace wider than long
- Two-jointed anal tubercle with a double fringe of curved setae
- Posterior spinnerets two-segmented, with distal segment long and curved
- Cribellate and ecribellate genera

on the apical joint of the anal tubercle. The circling behavior is carried out with the spider facing its prey, switching direction, and wagging the abdomen back and forth as it goes. This unusual type of attack is shared with the closest relatives of the oecobiids, the hersiliids. Male *Oecobius* build a special web to copulate with the female.

ABOVE | *Uroctea durandi*, from the Mediterranean region, lives under rocks in an upside-down silk tent with protruding edges that help the spider detect approaching prey.

HERSILIIDAE: *HERSILIA*

TWO-TAILED SPIDERS

The most striking features of the almost 80 described *Hersilia* species are their long spinnerets, reminding us that these were leglike appendages in the evolutionary past. In most spiders they are reduced to short structures, but in the two-tailed spiders, the posterior lateral spinnerets are strongly elongated—they can even exceed the length of the spider's cephalothorax and abdomen combined. These spinnerets are equipped with a long row of spigots (small, hollow, seta-like structures that are connected via narrow tubes with the silk glands). Using the spinnerets, *Hersilia* can produce broad bands of fine silk strands. These are used in their special technique of prey wrapping. *Hersilia* spiders live on tree bark or other smooth surfaces such as rocks, where they wait in a typical position for prey. If an insect passes, the spider runs lightning quick in circles around the insect, sometimes changing direction rapidly. During this process it produces from its spinnerets a silver band of threads that wraps the prey and

LEFT | A female *Hersilia* from Singapore, showing the eye arrangement on an anterior tubercle.

RIGHT | *Hersilia sumatrana* resting on a tree trunk in Sabah, Borneo. Note the elongated spinnerets and the excellent camouflage provided by the coloration pattern.

GENUS
Hersilia

DISTRIBUTION
Tropical and subtropical Africa and Asia, Australia (Western Australia, Northern Territory)

HABITAT
On bark of trees or rock surfaces, mainly in forests

CHARACTERISTICS
- Long spinnerets
- Special prey-wrapping technique
- Eight eyes on a large tubercle

immobilizes it against the tree trunk. The spider's egg sacs are attached to a tree trunk and guarded by the females.

The other 15 genera assigned to the family are similar in habits and appearance, although in some the posterior lateral spinnerets are not quite so elongated. Most of the family's diversity occurs in Africa, Eurasia, and Australia, whereas in the Americas only a few species are known so far. Hersiliids are considered close relatives of the Oecobiidae—members of which have less strikingly elongated spinnerets, yet capture their prey in much the same fashion.

DEINOPIDAE: *DEINOPIS*

OGRE-FACED SPIDERS

BELOW | An ogre-faced spider from Madagascar holding its catching web, ready to cast it around an insect flying by.

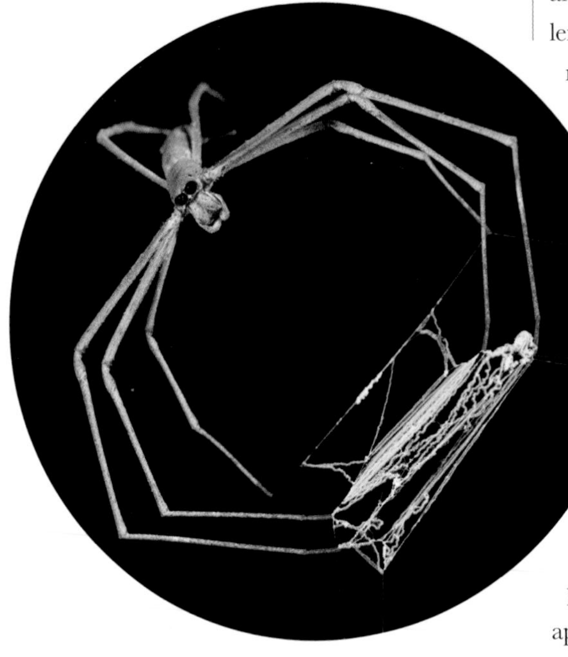

Two genera of deinopids are currently recognized. *Deinopis* occurs from Mexico, Florida, and Cuba to Argentina, throughout tropical Africa and Madagascar, and from Southeast Asia to South Australia. Their eight eyes are situated in three rows, with the posterior median pair greatly enlarged, giving them an ogre-like appearance when viewed from the front. The enlarged eyes, which may be the largest simple eyes relative to body size of any invertebrate, give the animal superb night vision; they have wide-angle lenses that proportionately capture 2,000 times more light that a human eye. Remarkably, the photosensitive retina is mostly destroyed every day at dawn, and then resynthesized at dusk.

Deinopids' front legs are slender and long—they can extend to more than three times their body length. Their closest relatives seem to be the uloborid orb-weavers. Rather than constructing a typical orb web, however, they spin only a few frame lines on which they hang from their back legs. Using the front four legs, they then grasp the corners of a rectangle consisting of several closely spaced, parallel lines of sticky silk. When a prey item approaches, the spider drops toward it. The

GENUS
Deinopis

DISTRIBUTION
Tropics and subtropics

HABITAT
Bushes and shrubs

CHARACTERISTICS
- Enlarged posterior median eyes
- Superb night vision
- Prey-catching web held in front legs

catching web is quickly expanded and cast around the prey, enveloping and immobilizing it for feeding. Specimens seem to be rare, at least in collections; relatively few adults have been taken, possibly because of insufficient collecting at night. The second genus in the family, *Menneus*, occurs in Australia, New Caledonia, southern and eastern Africa, and Madagascar.

ABOVE | An anterior view of an ogre-faced spider from Madagascar; note the greatly enlarged posterior median eyes, which help create the "ogre-like" appearance.

CRIBELLATE ORB-WEAVERS

Uloborids are highly unusual in many respects. Most, like *Philoponella*, construct orb webs—the regular geometric structures for which araneid spiders are famous. Unlike typical orb webs, however, uloborid webs are constructed using cribellate silk rather than sticky silk. In contrast to almost all other spiders, uloborids lack cheliceral venom glands, and thus have a very different mode of feeding. Whereas most spiders use their fangs to inject venom that helps to paralyze their prey, the mouthparts of uloborids usually never touch the prey during capture. Instead, these spiders rely on wrapping their prey in voluminous amounts of silk (up to several hundred yards of silk lines per prey item). The time spent wrapping prey is about two orders of magnitude longer than in araneid orb-weavers.

After the prey is completely wrapped and thus immobilized, uloborids then adopt a characteristic feeding position. The prey package is transferred from the legs to the palps and chelicerae, the anterior legs are spread apart, and the spider then rapidly rotates the prey. As it does so, it moistens the entire surface of the wrapped package with digestive

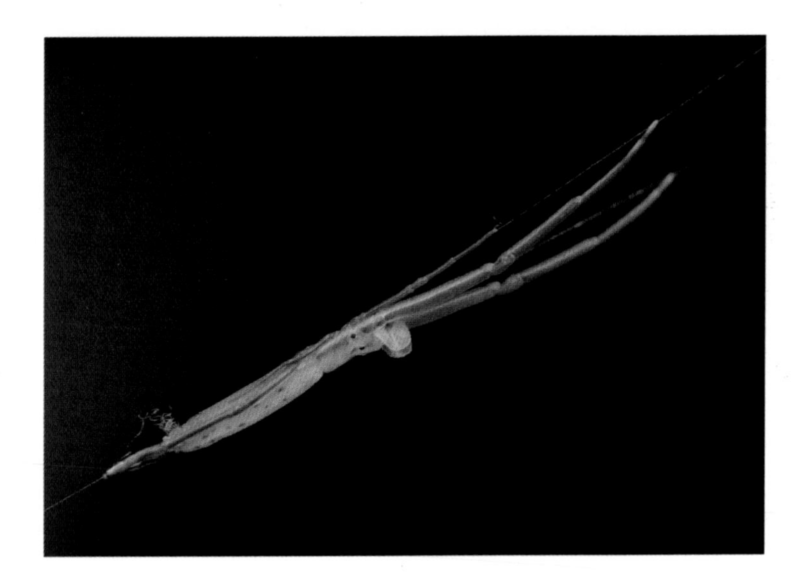

RIGHT | Subadult male *Miagrammopes* from Singapore. Note that the spider holds two threads at its front and one thread on its back, building a living bridge between the silk strands.

GENUS
Philoponella

DISTRIBUTION
Tropics and subtropics

HABITAT
Vegetation in forests and gardens

CHARACTERISTICS
- Cribellate orb web
- Absence of venom glands
- Femora with rows of long trichobothria

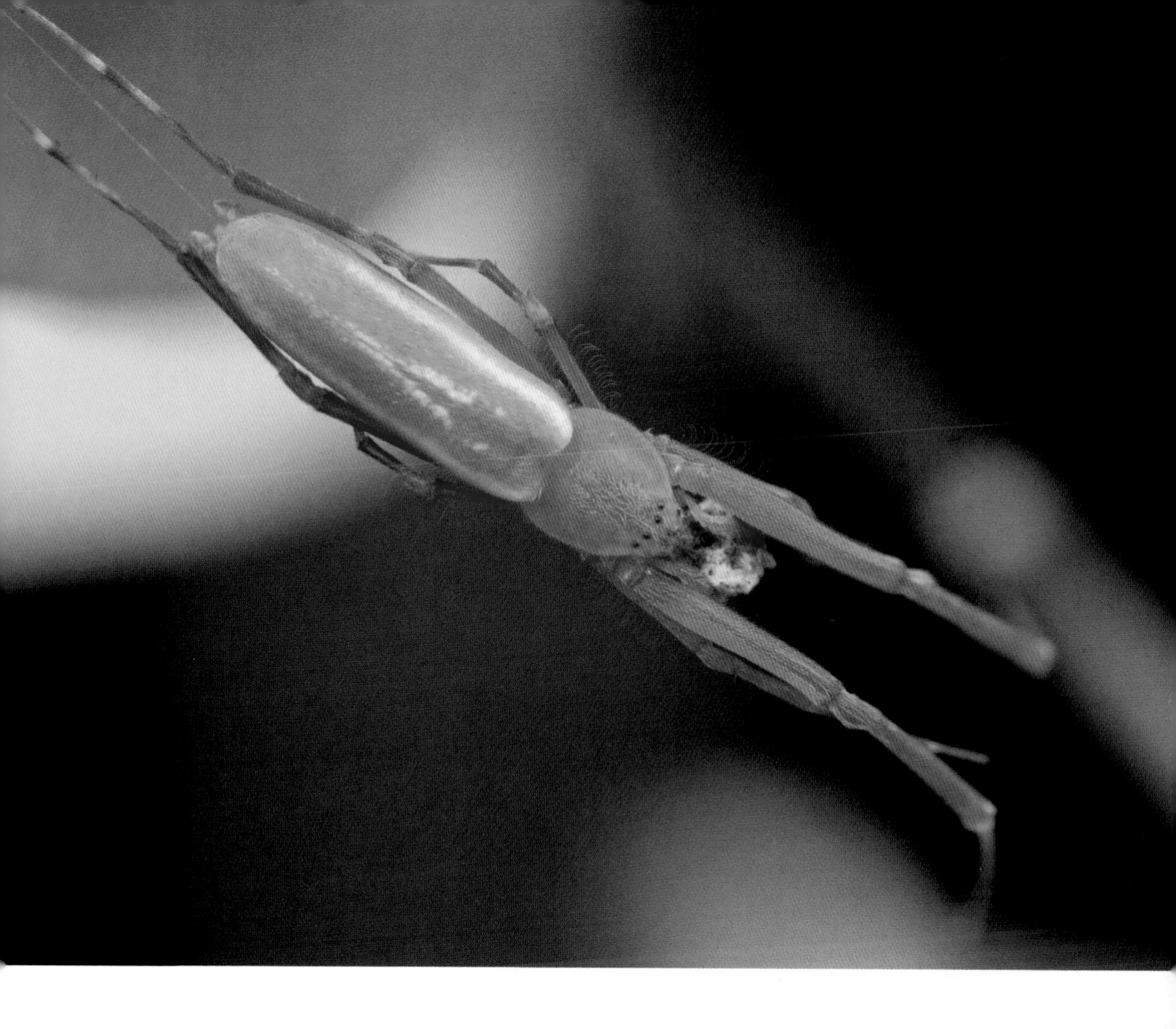

enzymes regurgitated from its midgut. The shroud of white silk soon becomes almost transparent, as highly effective enzymes proceed to digest much of the silk and its contents. Although the prey's cuticle can be damaged during the wrapping process, the enzymes can apparently digest the membranes of an insect's body and thereby enter the prey's body to liquefy all its contents, as in all other spiders. To avoid the action of its own enzymes, the spider typically touches the prey package only with the claws and thick setae at the tips of its palps. Some *Philoponella* species are known for building communal webs, i.e. many orb webs are built together, producing a huge barrier for flying insects.

Some other genera in the family build highly modified orb webs, showing at least partial omissions. In *Hyptiotes*, only three sectors of an orb are built: these are then pulled taut by a silken thread held in the leg claws of the spider. In *Miagrammopes* only one or two lines are built and act as capture threads. These spiders often have bodies modified for camouflage in vegetation, blending in with thin twigs. In many uloborids, the legs look feathery, with bunches of long bristles helping to disguise the spider's appearance. Similar puffs of setae may also be present on the abdomen.

ABOVE | A female *Philoponella* digesting a prey item. The legs are sometimes spread even wider when holding larger prey. Note the rows of long, fine trichobothria on the femora, used to detect airborne vibrations.

ARCHAEIDAE: *ERIAUCHENIUS*

ASSASSIN SPIDERS

Assassin spiders (sometimes also called pelican spiders) comprise more than 90 extant described species in five genera and have a Gondwanan distribution, including Australia, South Africa, and Madagascar. Archaeids have a substantial fossil record, dating back to the Jurassic period. In fact, the family was first described from Baltic amber fossil specimens in the mid-nineteenth

BELOW | *Eriauchenius workmani* from Madagascar with its impaled theridiid spider prey, displaying its highly maneuverable and elongated chelicerae. The carapace wraps around the base of the chelicerae, forming a long, narrow neck.

GENUS
Eriauchenius

DISTRIBUTION
Madagascar

HABITAT
Forest floor, among vegetation

CHARACTERISTICS
- Cephalic area highly elevated, forming a tubular "neck"
- Long, highly maneuverable chelicerae in which the two paturons can be moved independently
- Cheliceral peg teeth, and cheliceral and pedicel stridulatory organs

century, before the discovery of the first living representatives in Madagascar in 1881. The most salient morphological feature of archaeids is their elevated cephalic area, which forms a "neck," combined with their highly maneuverable, elongated chelicerae. The carapace and "neck" shape and elongation vary greatly among species, from short and stout to long and narrow.

Eriauchenius workmani, like all other archaeids, is exclusively araneophagic (i.e., it eats only other spiders) and does not build foraging webs. This Malagasy archaeid attacks its spider prey from a distance. Its extremely long chelicerae are held at 90° from the body, keeping the impaled spider prey safely away from the predator's own body until it is immobilized. The chelicerae and pedicel contain stridulatory organs—striae or ridges on the cuticle that produce sound when rubbed together. Archaeid species have narrow geographic distributions and can be found on the forest floor or on vegetation, where they ambush their spider prey.

ABOVE | A male *Eriauchenius* from Madagascar; note the rounded, orange palpal bulb held above the chelicerae.

MIMETIDAE: *ERO*

PIRATE SPIDERS

Mimetids are medium-sized to small spiders, their body rarely reaching more than 0.28 inches (7 mm) in length. About 150 species have been named, but many more remain to be described. Although mimetids are closely related to a lineage of ecribellate orb-weaving spiders, the Tetragnathidae (longjawed orb-weavers), they have abandoned capture webs, evolving instead a highly specialized foraging behavior which gives them their common name, "pirate spiders." *Ero*, together with most if not all other mimetids, prey primarily on other spiders, mainly araneids and theridiids. They invade the webs of their prey and mimic the behavior of a struggling insect or a potential mate. With slow and stealthy movements, the attacker inches closer to its prey. When it is near enough it

RIGHT | A male *Ero tuberculata*, a Palearctic pirate spider, photographed in France.

GENUS
Ero

DISTRIBUTION
Cosmopolitan, except Australia and New Zealand, although a common Palearctic species has been introduced to Australia

HABITAT
Usually found on vegetation (shrubs, trees, etc.)

CHARACTERISTICS
- Legs I and II are long, with a characteristic row of spines on the tibiae and metatarsi, consisting of long spines intercalated by shorter ones
- Chelicerae long, fused at the base and projecting vertically, with peg teeth on the promargin
- Ecribellate, not spinning foraging webs
- Preys mainly on other spiders

uses long forelegs bearing modified macrosetae to grasp the prey spider; it then bites one of its legs. Mimetids' strong venom kills prey very quickly, enabling feeding to begin. In some cases, the eggs of the prey are also eaten.

Ero species build characteristic egg sacs, with a stalk and wiry silk covering the sac itself. These are then laid on vegetation and abandoned by the mother. However, a few species of pirate spiders do show maternal care. Some neotropical species of *Mimetus* build nursery webs under a leaf; here the female will guard the eggs and the new emerged spiderlings. Females in the African genus *Anansi* are known to transport their spiderlings in their chelicerae. *Gelanor* males are unique among mimetids in having unusually long pedipalps, about twice the length of their body. Almost nothing is known about the biology of these beautiful spiders. *Gelanor* is the only mimetid genus exclusively found in the neotropics.

ABOVE | A male *Gelanor latus*, from the Rio Negro, Brazil. The adult males of *Gelanor* have extremely long pedipalps, with an unusual elongation of the femur, patella, and tibia.

MALKARIDAE: *MALKARA*

PITTED SHIELD SPIDERS

Malkarids are small to very small, cryptic, entelegyne spiders. They are found mainly in Australia and New Zealand: only two of the 46 described species so far have been found outside that region (*Flavarchaea humboldti* from New Caledonia, and *Chilenodes australis* from Chile and Argentina). Many species and several new genera remain to be described, and very little is known about their biology. Malkarids live in leaf litter and in mosses of wet forests, but some species are arboreal, living under tree bark or among dense foliage. They do not build foraging webs. Many malkarids, particularly some of the New Zealand species, have strong spination on the first two pairs of legs, resembling that of pirate spiders (Mimetidae), with alternating long and short spines. Malkarids can be distinguished from mimetids by the unique orientation of the conductor in the male palp. In some genera, the sternum is fully fused to the carapace.

BELOW | The female of an undescribed *Malkara* species from Queensland, Australia, found in the leaf litter of wet forests.

GENUS
Malkara

DISTRIBUTION
Australia and New Zealand (other genera in New Caledonia and Chile)

HABITAT
Among leaf litter and mosses in damp, humid forests; in trees, under bark or among foliage

CHARACTERISTICS
- Sclerotized ring around spinnerets
- Setae arising from sclerotized disks on abdomen
- Posterior lateral spinnerets lack the spigots involved in the production of viscid sticky silk (aggregate and flagelliform silk-gland spigots)

TRIANGLE SPIDERS

The family Arkyidae includes two genera of Australasian spiders, *Arkys* and *Demadiana*, with 37 described species. Some *Arkys* species are brightly colored (red, orange, yellow, and black) and have a distinctive triangular abdomen; other species mimic bird droppings. Both sexes of *Arkys* have large and powerful legs I and II, with thick spines used to capture their prey. Arkyids do not build foraging webs. Instead they can be found on vegetation, waiting to ambush their prey.

BELOW | A female *Arkys* from Australia, displaying the typical strong spination of the first two pairs of legs.

GENUS
Arkys

DISTRIBUTION
Australia, New Guinea, and Buru Island (Indonesia)

HABITAT
On vegetation—shrubs, trees, etc.

CHARACTERISTICS
- Prolateral field of short, dense setae on tarsus I of males
- The posterior lateral spinnerets of both sexes have enlarged trumpet-shaped spigots, used to make the sticky silk glue (aggregate silk-gland spigots), and lack the spigots that produce the cord fibers of the viscid sticky silk (flagelliform silk-gland spigots)
- Procurved posterior eye row—posterior median eyes positioned anterior to the posterial lateral eyes

ANAPIDAE: *ANAPIS*

GROUND ORB-WEB WEAVERS

Anapids are minute orb-weaving spiders with a worldwide distribution, although they are more diverse in the tropics and in southern temperate forests. More than 200 species have been described, classified in 58 genera, but their cryptic lifestyle and often narrow distribution ranges suggest that many species still await discovery. Most anapids build their small webs close to the ground, in moist environments, often in leaf litter. It is common to find several anapid webs very close to one another.

The web architecture of anapids is quite variable. Some species build sheet or irregular tangle webs, but many species build horizontal orb webs with out-of-plane radii that are pulled into a cone above the web. At least one anapid species can live as a kleptoparasite in the webs of other spiders (e.g., *Sofanapis antillanca*, from Chile, is often found in the sheet webs of austrochilids). *Anapis* includes close to 30 described species, all from the neotropical region. Probably all *Anapis* build orb webs with out-of-plane radii, but nothing is known about the biology of most species.

BELOW | Side and front views of an anapid from Borneo; the elevated clypeus and head region, and the high, triangular abdomen with platelets, are typical of anapids.

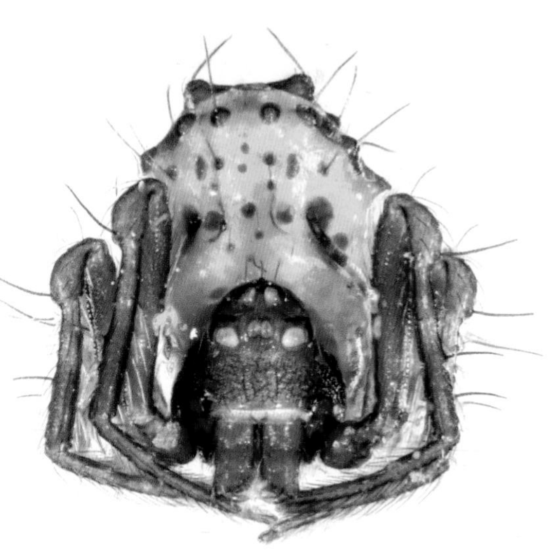

GENUS
Anapis

DISTRIBUTION
Mexico to Peru and Brazil

HABITAT
Close to ground in moist environments; often in leaf litter

CHARACTERISTICS

- Small to minute ecribellate spiders, smaller than 0.08 inches (2 mm), with eight or six eyes
- Female pedipalp absent or reduced in many species
- Labium fused to the sternum
- Carapace with pore-bearing prosomal depressions
- Abdomen with conspicuous sigilla (circular impressions or dimples) and with scattered, sclerotized spots
- Female genitalia entelegyne or secondarily haplogyne

ABOVE | The small orb web of a *Maxanapis* spider from New South Wales, Australia, built in the leaf litter. This architecture is typical of the webs of many anapids, with out-of-plane radii above the web and numerous spiral turns. The spider is usually found under the hub, upside down.

FALSE WIDOW SPIDERS

Steatoda belongs to the gumfoot spiders (Theridiidae), also called combfoot spiders—one of the largest spider families. Although the family already includes more than 2,500 described species, this number is expected to grow considerably because theridiids are abundant in tree canopies, a habitat still very poorly explored.

The genus *Steatoda* includes around 125 species. They occur in a wide variety of habitats, ranging from semidesert to wet forest, high mountains and even houses, where they are among the most commonly encountered spiders.

Males of *Steatoda* are easily recognized by the presence of a toothed ridge on the front of the abdomen, the pick of the stridulating organ. The other parts of the organ, the stridulatory ridges, are on the posterior portion of the carapace. *Steatoda triangulosa* is a very common house spider, present in both Europe and America. It builds the typical gumfoot web, characteristic of many theridiids. The web is usually situated under a rigid structure, such as a cupboard; it is composed of a tangle of threads in which the owner sits, and from which lines run to the ground. These threads are under

LEFT | As in many other spiders, the main prey of *Steatoda grossa* is insects, even hard-bodied ones like this *Coptocephala*, a member of the shield bug family.

RIGHT | *Steatoda triangulosa*, one of the false widow spiders, is among the commonest house spiders in the northern hemisphere; it lives a hidden life under cupboards or other furniture that is rarely moved.

GENUS
Steatoda

DISTRIBUTION
Cosmopolitan

HABITAT
Semi-desert to wet forests (mostly close to the soil); also high mountains and buildings

CHARACTERISTICS

- Anterior median eyes larger than other eyes
- Tarsi IV has a comb of serrated, curved bristles
- Males have a stridulating organ on anterior margin of abdomen

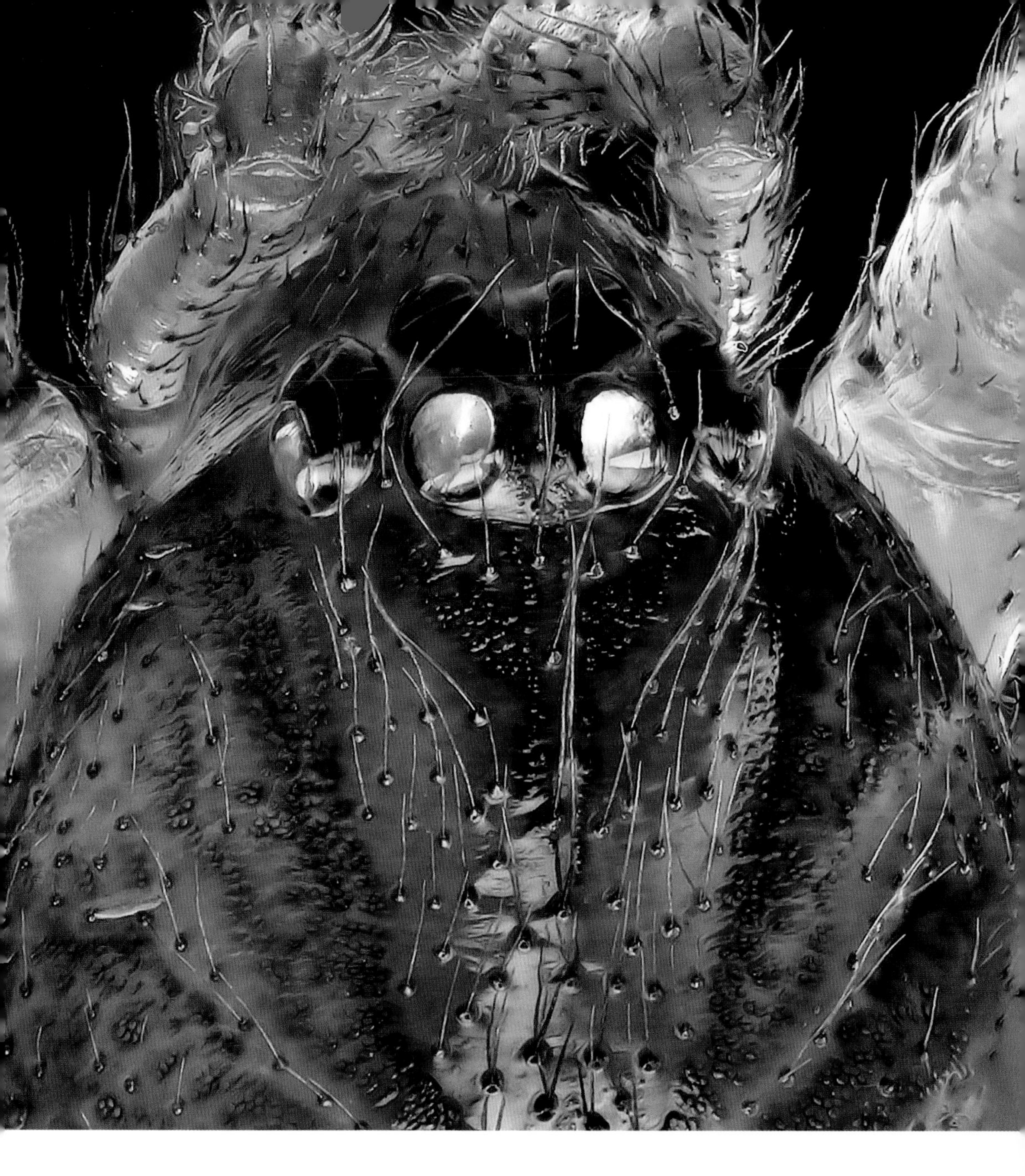

tension and have sticky drops near the end—the so-called "gumfoot." When a small prey animal touches the gluey drops, it gets stuck; the spider than severs the thread, lifting it from the ground and reeling in the prey.

Steatoda bear a superficial resemblance to widow spiders (*Latrodectus*), and are therefore sometimes called "false widows." They are usually smaller, with males and females of a similar size, and they always have venom that is far less potent to humans, although a bite may still be painful.

THERIDIIDAE: *EPISINUS*

CABLE SPIDERS

More than 60 species have been described in the genus *Episinus*, known as cable spiders. Their general appearance is characterized by the truncate abdomen and by short legs III. Males are usually somewhat smaller than females and have a less strongly truncate abdomen. Courtship and copulation are simple and straightforward, lasting only a few minutes.

These spiders construct an ingenious web that almost exclusively catches ground-dwelling animals, although some species prefer higher strata in the vegetation. This surprisingly simple web has the shape of an "H" or an inverted "Y." The top part is attached to a rigid support and also carries the spider. The lower part is held by the front legs of the owner and consists of only two strong threads, the "cables," provided with gluey drops near the extremity, as in many other theridiid webs. These gumfoot lines immobilize any prey that touches them and it is then overwhelmed by the spider.

LEFT | Their color pattern and unusually truncated abdomen make cable spiders (*Episinus*) less conspicuous when they hang head down in their simple web.

RIGHT | The long first and fourth leg pairs are clearly adaptations to the peculiar strategy of cable spiders like this European *Episinus angulatus*. The spider hangs from the crosspiece of the H-shaped web and keeps its front legs along the longitudinal threads.

GENUS
Episinus

DISTRIBUTION
Cosmopolitan

HABITAT
On the ground or in vegetation

CHARACTERISTICS
- Truncated abdomen
- Legs of third pair very short
- Holds gumfoot lines in H-shaped web

THERIDIIDAE: *LATRODECTUS*

WIDOW SPIDERS

Latrodectus includes more than 30 species, some of which are among the most infamous venomous animals. The black widows—*Latrodectus mactans* in America and *L. hasselti* in Australia—not only have potent, neurotoxic venom, but are also frequently found in or around habitation. Efficient antivenoms are readily accessible in these high-risk areas. Other widow spiders, such as the Brown Widow (*L. geometricus*) and the Madagascar Widow (*L. menavodi*), either have venom with lower toxicity or are rarely found in man-made habitats.

Widow spiders have earned their name because the males are much smaller than the females and are almost always devoured after copulation. Sexual cannibalism is not rare among spiders: males often have longer legs or use other

LEFT | The Redback Spider (*Latrodectus hasselti*), common in Australia, is easily recognized by the dorsal abdominal pattern with its conspicuous red patch; it is infamous for its potent neurotoxic venom.

GENUS
Latrodectus

DISTRIBUTION
Tropical and subtropical areas worldwide; southern Palearctic

HABITAT
Synanthropic; more abundant in disturbed than natural habitats

CHARACTERISTICS
- Lateral eyes clearly separated
- Males much smaller than females
- Embolus (male copulatory organ) long and coiled
- Often with a red hourglass pattern on the abdominal venter

strategies to stay out of reach in case the female they are courting is more interested in a meal than in copulation. But the males of several *Latrodectus* species somersault over 180° during mating, landing on the mouthparts of the female they have inseminated. By offering themselves as prey, the males provide resources for the development of the eggs, which number in the hundreds per egg sac. The sacs are hung in the female's tridimensional web, which is made of very tough silk threads.

A female that has copulated and eaten a male is apparently less attractive to other males. Nevertheless, males also possess other strategies to render their partners less accessible to rival males. Often, part of the emboli (the male's copulatory organs) breaks off during copulation; it remains in the female genital tract, thus impeding the insertion of additional emboli. Before copulation, males often also reduce the female's web by excising and fusing some of the threads; they cover these with their own silk, which is impregnated with male contact pheromones, making it more difficult for other males to detect the female.

ABOVE | The Madagascar Widow (*Latrodectus menavodi*) shows the red hourglass spot on the ventral side of the abdomen, as is typical of most widow spiders. It is a rare species restricted to peculiar habitats, but it shares the potent venom of other widows.

MYSMENIDAE: *MYSMENA*

MINUTE CLASPING WEAVERS

BELOW | A female *Mysmena tasmaniae*, from Tasmania. The adult female measures about 0.04 inches (1 mm) and builds its web in the forest litter, in mosses, and at the base of grass tussocks.

Mysmenids are small orb-weavers, usually less than 0.08 inches (2 mm) in length. The family has a worldwide distribution. They live close to the forest floor, in leaf litter and small ground cavities, in humid environments, and are most diverse in the tropics. Some mysmenids, such as *Mysmenopsis*, are kleptoparasites; they live in the webs of other spiders and feed on tiny prey that may be ignored by the larger spider. Although about 130 species have been described, many known from collections still remain undescribed. *Mysmena* includes about 50 described species that build a spherical, tridimensional orb web, with a proliferation of out-of-plane radii and sticky spirals both above and below the orb plane. Other mysmenids build webs similar to those of some anapids (*Maymena*) or sheet webs (*Trogloneta*).

LEFT | The highly modified, tridimensional orb web of a *Mysmena* species from Madagascar.

GENUS
Mysmena

DISTRIBUTION
Holarctic, South America, Australasia, Pacific islands

HABITAT
Humid forests, especially in the tropics; close to the forest floor, in leaf litter or small cavities

CHARACTERISTICS
- Femoral spot (a sclerotized area subapically on ventral femora I and II)
- Male metatarsus I with a clasping spine
- Male palp with an apical cymbial conductor
- Respiratory system with anterior tracheae restricted to opisthosoma (abdomen)

SYMPHYTOGNATHIDAE: *PATU*

DWARF ORB-WEAVERS

The family Symphytognathidae includes the smallest adult female spider—*Anapistula ataecina*, from Portugal, with a body length of 0.017 inches (0.43 mm)—and the smallest adult male spider—*Patu digua*, from Colombia, with a total length of about 0.015 inches (0.37 mm). About 70 symphytognathid species have been described, in eight genera, mostly from tropical regions. They live in very humid environments, building their webs in leaf litter and small cavities in the forest floor or fallen logs. They weave dense, planar, horizontal orb webs, extraordinarily large relative to the tiny size of the spider. These webs have many radii (most of which are accessory radii, laid after sticky spiral construction) and spiral turns. Symphytognathids keep their egg sacs at the margin of their webs.

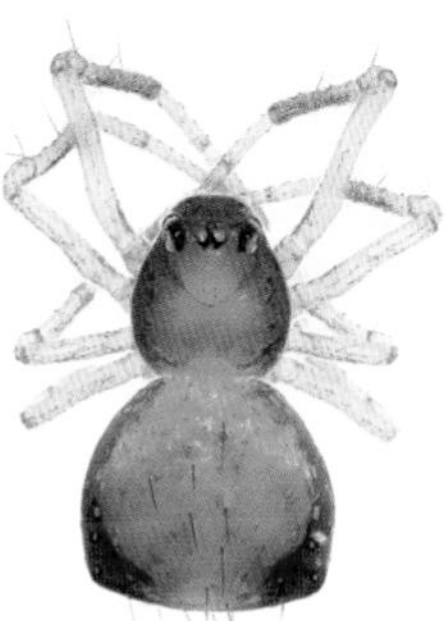

ABOVE | A female *Patu quadriventris* from China; the total length of this animal is only 0.03 inches (0.75 mm)—about the size of the period at the end of this sentence!

LEFT | The planar orb web of *Patu* in the leaf litter of a forest in the Dominican Republic.

GENUS
Patu

DISTRIBUTION
Tropical regions, China, Japan

HABITAT
Humid forest environments; on the forest floor in leaf litter, fallen logs, or ground cavities

CHARACTERISTICS
- Very small size, less than 0.03 inches (0.7 mm)
- Six eyes in three diads and an elevated pars cephalica (head region)
- Chelicerae fused near their base
- Female palps absent
- Book lungs replaced by tracheae

TETRAGNATHIDAE: *TETRAGNATHA*

LONG-JAWED ORB-WEAVERS

Tetragnathidae is worldwide in distribution, but the family is most diverse in the tropics. It includes more than 1,000 described species in about 50 genera. Most tetragnathids build orb webs, which are frequently oriented horizontally and have an open hub, but there is much variation in web orientation, the number of radii, and the number of spiral turns.

Tetragnatha is the largest genus in the family, with more than 350 species described so far. In the Hawaiian archipelago, *Tetragnatha* has undergone a spectacular adaptive radiation, resulting in about 60 species that differ widely in their ecology, habitat preference, webs, coloration, shape, and size. Most continental *Tetragnatha* species build flimsy horizontal webs with few radii and spiral turns. These are often close to the water surface of ponds and streams, where nematocerous flies and other small, delicate flying insects abound.

LEFT | A female *Tetragnatha*, displaying its characteristic elongated body and long, thin legs.

GENUS
Tetragnatha

DISTRIBUTION
Cosmopolitan

HABITAT
Webs on vegetation or close to the ground, often near ponds and streams

CHARACTERISTICS

- Male palps lacking a median apophysis and with conductor wrapping and coiling with the embolus
- Many species have long chelicerae with strong teeth (in *Tetragnatha*)
- Fourth femora (femora IV), sometimes with trichobothria
- Tetragnathine female genitalia are secondarily haplogyne

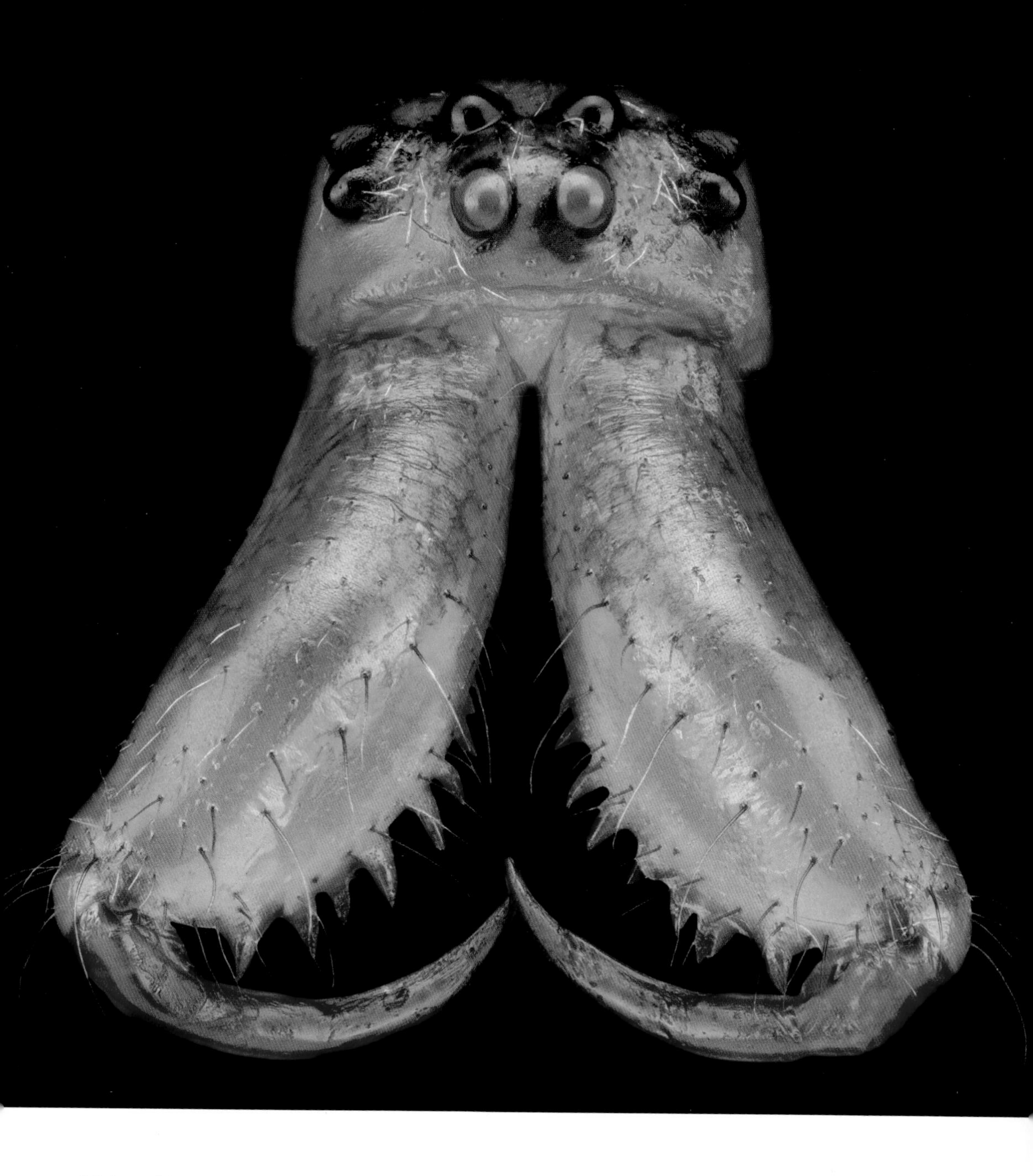

Male and female *Tetragnatha* interlock their chelicerae during mating. The males have paturons with special projections or spurs that interact with the female chelicerae in such a way that, when clasped, the females cannot bite them. The female will then curve her abdomen to bring it closer to the male's, and in this position the male will insert his palps alternately.

ABOVE | *Tetragnatha* species, like this *T. montana* female, have long and robust chelicera with numerous teeth. The number, size, and arrangement of cheliceral teeth varies among species.

SYNAPHRIDAE: *CEPHEIA*

MIDGET SPIDERS

Synaphrids are tiny, araneoid spiders that live in southern Europe, the Canary Islands, the Mediterranean region, central Asia, and Madagascar. Very little is known about the biology and natural history of synaphrids, and only about a dozen species have been described, in three genera. *Cepheia longiseta*, the only known species in its genus, is found in southern Europe, where it has been collected in coastal regions, spinning tiny webs in the ground vegetation.

Like other synaphrids, *Cepheia* has its anterior book lungs reduced to tracheae, with five tracheal tubes arising from each anterior spiracle. The posterior tracheal system has two distant spiracular openings, connected externally by a thin ridge. Both tracheal systems seem to reach into the prosoma.

RIGHT | A female *Cepheia longiseta*, the only species in its genus. They live in the western Mediterranean region and their adult body size is less than 0.04 inches (1 mm) long.

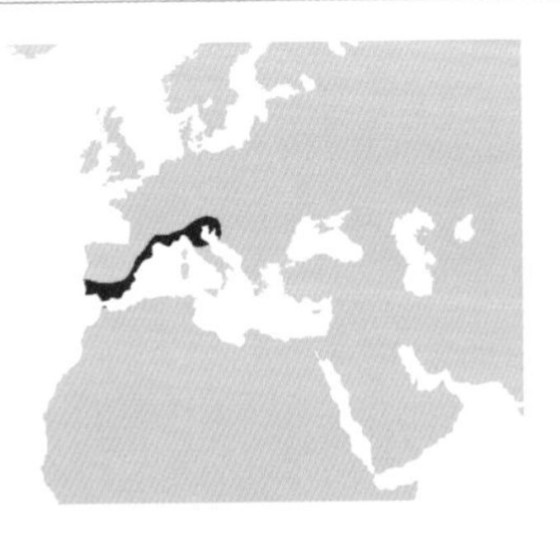

GENUS
Cepheia

DISTRIBUTION
Southern Europe

HABITAT
Coastal regions, in vegetation close to the ground

CHARACTERISTICS

- Tarsi–metatarsi joint constricted on every leg in both sexes; no leg spines
- Chelicerae with median keel ending in single strong prolateral tooth; no retromarginal teeth
- Carapace rounded, as long as wide, with the clypeal area (between the anterior margin of the carapace and the anterior eyes) protruding in dorsal view
- Male palp compressed laterally, very large, almost as large as prosoma; a distinctive tibial morphology, rounded retrolaterally and pressed against the cymbial base; female palp clawless

CYATHOLIPIDAE: *FORSTERA*

TREE SHEET-WEB SPIDERS

BELOW | An adult male *Teemenaarus* from Queensland, Australia (a close relative of *Forstera*) displaying its characteristic elongated triangular abdomen with long setae on its tip, thickened and curved first femur, and sinuate first metatarsus.

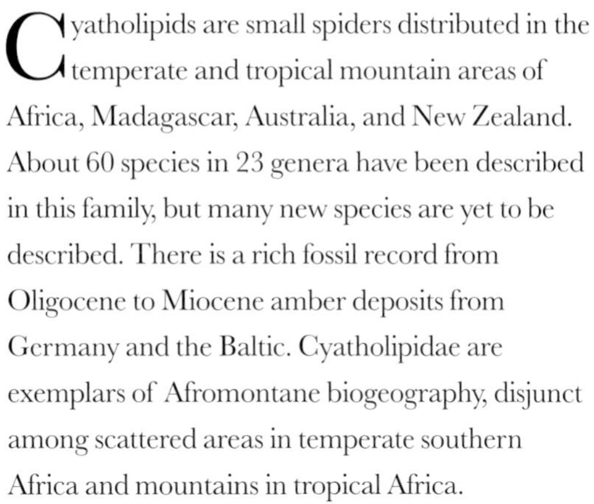

Cyatholipids are small spiders distributed in the temperate and tropical mountain areas of Africa, Madagascar, Australia, and New Zealand. About 60 species in 23 genera have been described in this family, but many new species are yet to be described. There is a rich fossil record from Oligocene to Miocene amber deposits from Germany and the Baltic. Cyatholipidae are exemplars of Afromontane biogeography, disjunct among scattered areas in temperate southern Africa and mountains in tropical Africa.

Forstera, like other cyatholipids, build horizontal sheet webs on vegetation and tree trunks in humid forests, hanging upside down under the web. The web is densely woven and very large relative to the spiders' tiny size. Adult males retain the araneoid spigot triplet on the posterior lateral spinnerets.

GENUS
Forstera

DISTRIBUTION
Queensland, Australia

HABITAT
Temperate and tropical mountain areas; vegetation and trunks of humid forests

CHARACTERISTICS
- Males with sinuate metatarsus I and thickened, curved femur I
- Male palp with a laterally expanded cymbium, a cup-like paracymbium, and a unique cymbial retromedian process; no conductor
- Anterolateral spur on male chelicera
- Abdomen triangular when viewed laterally
- Posterior spiracles connected by a broad, transverse slit that is sclerotized at each end, and with branched tracheoles that enter the prosoma

PIMOIDAE: *PIMOA*

LARGE HAMMOCK-WEB SPIDERS

GENUS
Pimoa

DISTRIBUTION
Western North America, southern Europe, and Asia

HABITAT
Temperate forests, close to the forest floor in fallen trees, ground cavities, and among exposed roots; some species in leaf litter of mixed conifer forests in western USA; European and Asian species often in humid caves; possibly also synanthropic

CHARACTERISTICS
- Most species have cheliceral stridulatory striae
- Legs break at the patella–tibia joint (a point of weakness)
- Male palp with a retrolateral cymbial sclerite and a cymbial process bearing cuspules (short, modified setae)

*P**imoa* species are relatively common in the forests of western North America. The European species, as well as some of the Asian species, are often found in caves, where they seem to occur opportunistically because of the higher humidity. Most pimoids build flat sheet webs, usually close to the ground, and attached to fallen trees and stumps, among exposed roots, or in hollow cavities. The common North American species *Pimoa altioculata* can be found in buildings, other man-made structures, and debris in or near forested areas.

Pimoids hide in retreats at the edge of their webs in the daytime, emerging only at night. Like linyphiids (the pimoids' closest relatives), they hang and walk on the undersurface of their sheet webs. Some pimoid webs are very large, with a surface area of up to 3.3 square feet (1 square meter).

The closest relative of the genus *Pimoa* is the enigmatic species *Nanoa enana*, which is only about 0.06 inches (1.5 mm) long—in contrast to the much larger *Pimoa* species, which are 0.2–0.47 inches (5–12 mm) long. *Nanoa enana* lives in the leaf litter of mixed conifer forests of northern California and southern Oregon, USA. Although its natural history remains unknown, the spider's silk-gland spigot morphology suggests it does not spin foraging webs.

LEFT | A female of *Pimoa cthulhu*, a large species that spins its web in hollow trunks and large ground cavities in some of the redwood forests of California, USA.

BELOW | A female *Pimoa breviata*, a common species in the Pacific Northwest of the USA.

LINYPHIIDAE: *LINYPHIA*

EUROPEAN SHEET-WEB SPIDER

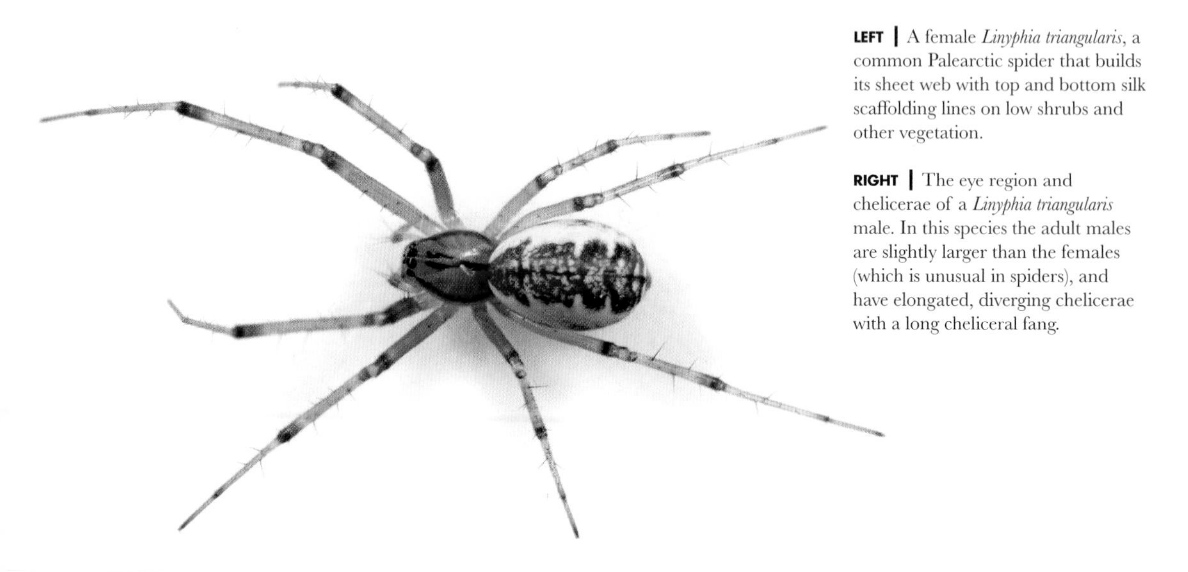

LEFT | A female *Linyphia triangularis*, a common Palearctic spider that builds its sheet web with top and bottom silk scaffolding lines on low shrubs and other vegetation.

RIGHT | The eye region and chelicerae of a *Linyphia triangularis* male. In this species the adult males are slightly larger than the females (which is unusual in spiders), and have elongated, diverging chelicerae with a long cheliceral fang.

Linyphiids are the most speciose family of araneoid spiders—a large clade that includes the ecribellate orb-weavers. Although linyphiids have a worldwide distribution, they are most diverse in north temperate regions. At higher latitudes, they account for a large fraction of the total spider diversity.

Linyphia triangularis is one of the most common European linyphiid spiders. It is found in a great diversity of habitats, building its sheet web on shrubs, tall plants, and trees. The web is slightly concave, with scaffolding lines both below and above the main sheet; those below are particularly dense. Like most linyphiids, *L. triangularis* is a generalist predator. It will feed on any insect that is intercepted by the upper scaffolding and falls onto the main sheet. All web-building linyphiids and their close relatives, the pimoids, hang upside down on the undersurface of their sheet web.

Linyphia triangularis is unusual among spiders in that adult males are often larger than adult females. After reaching the final molt, adult males move into a juvenile female's web, asserting dominance over the female and feeding on most of the prey that fall on

GENUS
Linyphia

DISTRIBUTION
Europe and Asia; introduced to the USA but not widespread

HABITAT
North temperate regions, in vegetation: shrubs, tall plants and trees

CHARACTERISTICS
- Stridulatory ridges on the side of the chelicerae
- Upside-down position below a horizontal web
- Legs long and slender; longest in males
- Chelicerae larger in males than in females
- Epigynum with a single opening, triangular, and with spiral folds

the web. Juvenile females emit a pheromone that attracts males. Males have larger chelicerae than females and they use them in aggressive interactions to try to expel all other potential suitors—usually larger males displace smaller ones from the female's web. While guarding the female, the male will damage and reduce her web; by this action he reduces evaporation of the male-attractant chemicals in her web, thus decreasing the chances of her attracting rival males. When the female becomes sexually mature, she copulates immediately with the male that has successfully evicted all his competitors. In *Linyphia* and its close relatives, the first male to mate with a female will sire the majority of her offspring.

LINYPHIIDAE: *LABULLA*

MONEY SPIDERS

The genus *Labulla* comprises three species of European spiders. *Labulla thoracica* is a common species, widely distributed in Europe, particularly in central and northern European countries. These spiders build sheet webs, mainly at ground level, in wooded areas. The webs are commonly found at the base of trees between exposed roots and buttresses, in overhanging banks, under logs, and in a variety of shady habitats, including man-made structures such as cellars and outhouses.

During the daytime the spiders hide in a retreat in the less exposed area of the web and are rarely seen; at night they are found upside down—a position typical of linyphiids and pimoids—in the center of the web. Adult males are often found together with females in their webs, suggesting that when males move into female webs they occupy them for some time after copulation. Their webs consist of a main horizontal platform or sheet, most often lacking any upper scaffolding—except perhaps for a few vertical silk lines, and a three-dimensional tangle of lines under the sheet. The web of *L. thoracica* traps small flying insects, such as fungus gnats, that seek shelter in the dark cavities of tree buttresses.

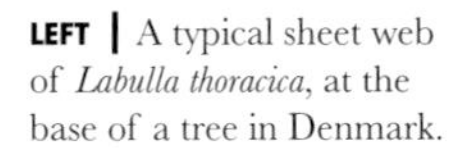

LEFT | A typical sheet web of *Labulla thoracica*, at the base of a tree in Denmark.

RIGHT | A female *Labulla thoracica*. Linyphiids walk upside down on the undersurface of their sheet webs.

GENUS
Labulla

DISTRIBUTION
Europe

HABITAT
Wooded areas, mainly at ground level near the base of trees; also shady areas, including cellars and outhouses

CHARACTERISTICS
- Medium-sized entelegyne spiders; females up to 0.27 inches (7 mm) long
- Robust chelicerae; long, annulated legs
- Adult male palps large, with a very long and filiform embolus forming a wide circle

LINYPHIIDAE: *WALCKENAERIA*

DWARF MONEY SPIDERS

Dwarf money spiders (subfamily: Erigoninae) comprise the largest group within the highly diverse and species-rich family Linyphiidae. Erigonines range in body size from 0.04 to 0.24 inches (1–6 mm), but most are about 0.079 inches (2 mm) in length (although there are some gigantic erigonine species in the genus *Laminacauda* in the Juan Fernández and Tristan Da Cunha Islands that reach 0.4 inches/10 mm). Erigonines are one of the dominant spider groups in the temperate and cold regions of the northern hemisphere. For example, in northern Europe there are more than 300 species of erigonines (about 25 percent of the total spider fauna of that area). Most erigonine species are leaf-litter dwellers and build tiny sheet webs.

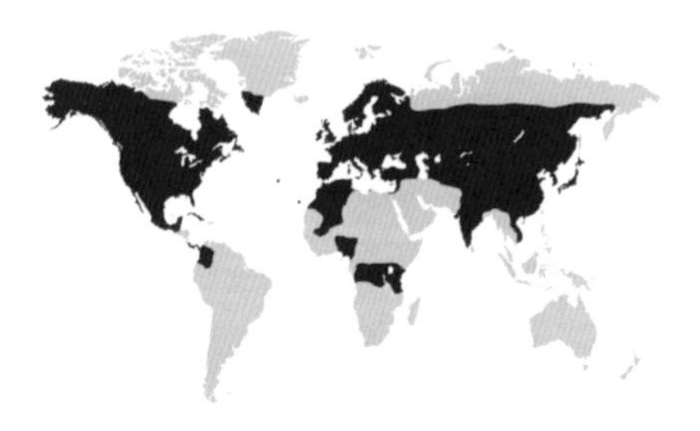

GENUS
Walckenaeria

DISTRIBUTION
The Americas, Europe, Africa, Asia

HABITAT
Temperate and cold areas; mostly in leaf litter

CHARACTERISTICS

- Small entelegyne spiders, usually less than 0.2 inches (5 mm)
- Male cephalic region often raised, with lobes, grooves, and modified setae
- Branched tracheal system; median trunks with tracheoles that enter the prosoma

ABOVE | A male dwarf money spider, *Laminacauda tuberosa*, from the Juan Fernández Islands.

LEFT | A male *Walckenaeria acuminata*, displaying its characteristic cephalic (head) region, which is projected into a turret that carries the eyes.

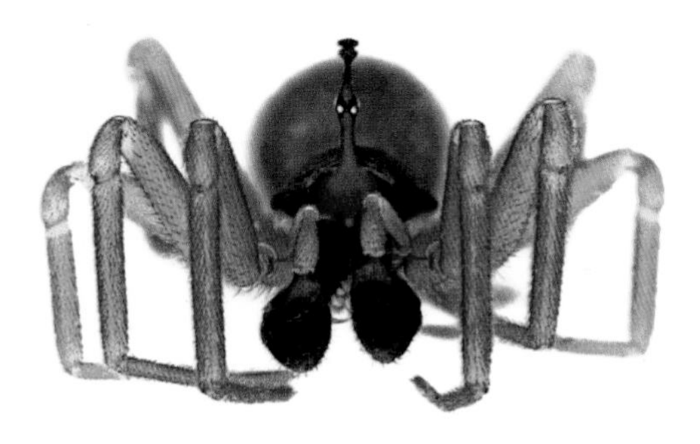

ABOVE | Anterior view of a male *Walckenaeria acuminata*. The females lack the cephalic turret.

In many species of dwarf spiders, such as those in the genus *Walckenaeria*, the males exhibit a vast morphological diversity of cephalic modifications, including lobes and turrets, grooves, pits, and modified setae. During mating, in the very few species that have been studied, the females use their cheliceral fangs to grip the pits or grooves on the male carapace during copulation. They then proceed to feed on secretions produced by specialized prosomic glands of the male, by depositing "saliva" on them and reingesting it.

Walckenaeria acuminata is a common European species that exhibits an extreme case of male cephalic structures. A high turret raised vertically over the carapace carries the eyes; this in turn is crowned by bundles of modified setae in the shape of tridents. The mating behavior of this species, like that of most erigonines, remains unknown.

SYNOTAXIDAE: *SYNOTAXUS*

CHICKEN-WIRE WEB SPIDERS

LEFT | A female *Synotaxus* from Brazil with her egg sac.

Synotaxus species live in the understory of wet forests in the American tropics. They build their capture webs between two leaves of trees and bushes. Their green coloration, and their habit of flattening their body against the underside of the leaf where they hide, make it difficult to find these spiders. Eleven species have been described so far, but it is likely that more remain to be discovered. *Synotaxus* species have a long abdomen (extending beyond the spinnerets to varying degrees), a wide and flat carapace, and long, delicate legs. Their legs and body are covered with long, fine, semi-erect setae.

Synotaxus are most easily characterized by the unique architecture of their vertical, planar webs, comprising one or several modules with a distinctive "chicken-wire" pattern. The sides of the web (and of the web modules) are formed by long, straight, non-sticky lines. The horizontal threads, which are roughly straight, are also made of non-sticky silk, while the jagged vertical threads are made of viscid, sticky silk with widely spaced glue globules. The capture threads of *Synotaxus* webs, unique among spiders, have a spring-like structure, presumably to allow great extensibility. Their webs are built (or rebuilt) early in the evening.

The sticky and non-sticky lines are laid in a very regular pattern, using a highly stereotyped behavior. In contrast with typical orb webs, such as tetragnathids and araneids, which organize their web building around a central hub, *Synotaxus* intersperses the laying of sticky and non-sticky lines without organizing the building activities around a central area. Once prey (usually small flying insects) is trapped in the web, *Synotaxus* use a silk wrap attack to subdue it, applying silk to the prey simultaneously with both fourth legs—a behavior similar to the prey-wrapping of theridiids and nesticids.

GENUS
Synotaxus

DISTRIBUTION
Costa Rica south to Paraguay

HABITAT
Understory of humid forests; webs are built in vegetation well above the forest floor

CHARACTERISTICS
- Vertical "chicken-wire" web, comprising modular rectangles with a regular pattern of sticky and non-sticky silk lines
- Enlarged (but not flattened) aggregate gland spigots on the posterior lateral spinnerets
- Male palp with a stout patellar spur

PHYSOGLENIDAE: *CHILEOTAXUS*

GREEN DOME-BUILDING SPIDERS

Physoglenids include 13 genera and 72 species with a partial Gondwanan distribution: they are found in the wet forests of Chile, Australia, and New Zealand. Additional genera and species are yet to be described. Most physoglenid species live in New Zealand and Australia, but the genera *Physoglenes* and *Chileotaxus* are found exclusively in Chile.

Chileotaxus sans, the only species in its genus, lives in central Chile. It builds a dome-shaped sheet web on shrubs, with its apex placed under a leaf. This long-legged spider is green in color and difficult to see against the background provided by the leaf, under which it sits upside down. Most other physoglenids make their irregular or sheet webs on the ground—some physoglenid webs are very similar to linyphiid webs.

BELOW | The web of a *Chileotaxus sans* in Chile. The spider, green in color, is barely visible under the leaf.

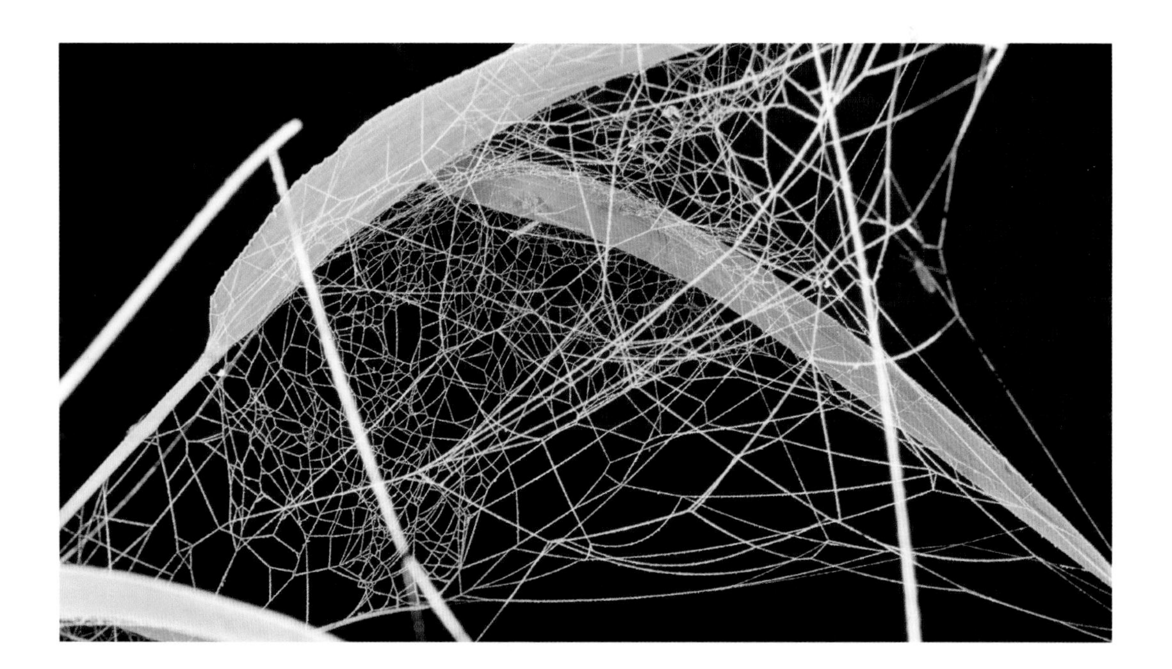

GENUS
Chileotaxus

DISTRIBUTION
Chile

HABITAT
Wet forests; mainly under the leaves of shrubs

CHARACTERISTICS
- Long, spineless legs with femora basally enlarged
- Male pedipalp with a small, dorsally excavated basal paracymbium and a retrolateral cymbial incision
- Cylindrical silk-gland spigots absent from the posterior median spinnerets; only one is present on the posterior lateral spinnerets

NESTICIDAE: *NESTICUS*

CAVE COBWEB SPIDERS

Nesticus is a widely distributed genus with nearly 125 species. The majority of these species are troglophiles, with caves as their preferred habitat—as are most of the other members of the family Nesticidae that occur in the north temperate parts of the world. One of the species, *Nesticus cellulanus*, is synanthropic; it has been introduced to North America, and is not restricted to caves, constructing cobwebs in sewage systems and other dark places. Females produce egg sacs with up to 50 eggs, which they carry attached to the spinnerets in a lycosid-like fashion. However, the spiderlings do not stay in the vicinity of the mother once they have hatched from the eggs.

Some mountainous areas of North America are home to numerous species. Each has a small distribution area, sometimes restricted to just one or a few caves. Since most males of these species have complicated palps, they are interesting for the study of the influence of isolation, sexual selection, and other possible mechanisms that bear on the evolution of the genitalic structures. Nesticids from tropical regions are not as frequently associated with caves, occurring also in leaf litter and under stones.

RIGHT | Although cave cobweb spiders are not rare, they are rarely seen since they live in dark sewage systems. *Nesticus cellulanus* is probably the most common species of the family as it is adapted to man-made environments. Females carry their egg sac attached to the spinnerets.

GENUS
Nesticus

DISTRIBUTION
Cosmopolitan

HABITAT
Mainly cave dwelling, especially in north temperate regions; in leaf litter and beneath stones in the tropics; also synanthropic (cellars and sewage systems)

CHARACTERISTICS
- Anterior median eyes small
- Tarsi IV with a comb of serrated curved bristles
- Spin a tridimensional web
- Often live in caves

ABOVE | The male *Nesticus cellulanus* has conspicuous palps with a huge outgrowth of the cymbium, the dorsal part of the copulatory organ. Such structures play a role in locking the intromittent organ, the embolus, in a specific position during copulation.

THERIDIOSOMATIDAE: *THERIDIOSOMA*

RAY SPIDERS

Theridiosoma contains some 30 species of tiny orb-weavers. Like other theridiosomatids, they have a group of long, vibration-sensitive trichobothria on the tibiae of both pairs of posterior legs. The anterior margin of the sternum is provided with a pit organ at both corners.

These spiders spin a remarkable, vertical orb web in damp, dark places. At the end of web construction, the radii at the center are fused, so that only four—the so-called rays—reach the hub. The spider's back legs hold these while the front legs clamp an extra line out of the plane, attached to nearby vegetation. Pulling on that line distorts the web into a cone and keeps it under tension. When possible prey nears the web, the spider releases the tension and the web, with its sticky spiral, slams onto the prey. The web owner then bites the victim and wraps it in silk.

The members of another genus of ray spiders, *Wendilgarda*, construct an even more unusual web. They spin threads that hang from a branch or cliff over slow-flowing water. These lines, provided with a few sticky drops, adhere to the water by surface tension, and catch prey floating on the surface.

Ray spiders tend to be difficult to detect, because of their relatively small size and their preferred habitats. However, their presence is often given away by their pretty, pale brown egg sacs. Each sac hangs from a single strand of silk, frequently in an exposed place, in contrast to the site of the web.

LEFT | Ray spiders (Theridiosomatidae) are among the smallest spiders. This explains why their study is still in its infancy. The number of species is expected to increase considerably with an exhaustive study of the many specimens that await description in museum collections.

GENUS
Theridiosoma

DISTRIBUTION
Cosmopolitan

HABITAT
Damp, dark places, usually with vegetation

CHARACTERISTICS
- Groups of long trichobothria on tibiae III and IV
- Sternum with pit organ on either side of anterior margin
- Cone-shaped orb web under tension
- Egg sacs hang from a thread

ABOVE | The web of a ray spider is an ingenuous construction.

LEFT | Though conspicuous, *Theridiosoma* egg cases are not easily preyed upon; predators have to be both small enough to penetrate in the confined spaces where the cases are hung and skillful enough to reach the eggs via the slender stalk.

ARANEIDAE: *ARANEUS*

GARDEN ORB-WEAVING SPIDER

The family Araneidae is very diverse, including more than 3,100 described species from all continents except Antarctica. Most, though not all, build orb webs, as do many species in other families (including tetragnathids, theridiosomatids, and uloborids). The genus *Araneus* contains more than 600 described species, but many of them are incorrectly placed, including all those species described from Australia.

Araneus diadematus, with a Holarctic distribution, is one of the most common araneids, especially in Europe. It builds a typical orb web in a variety of habitats, including gardens and orchards. Its orb web is a bidimensional and highly geometric vertical structure, with a well-defined frame; it encloses a set of radii converging on a hub and a sticky spiral.

There is a retreat usually above the web, hidden in leaves or under bark, which is connected to the hub by a signal line, so the spider can move quickly to the hub. Although most araneid species are nocturnal, hiding in the retreat at daytime and resting upside down at the hub at night, *A. diadematus* is also often found at its hub during the day.

Most araneids rebuild their webs daily (as they are usually damaged by prey or debris), using a highly stereotyped behavior; the old web is ingested and the proteins in the silk are recycled. A female araneid produces seven different types of silk, each used for a different purpose and with different physical properties. Males have an additional set of silk glands, with spigots near the epigastric furrow, that produce the silk used in the sperm web.

Some common araneids, such as *Cyrtophora* in the Old World and *Mecynogea* and its close relatives in the Americas, build modified orbs with a dome-shaped platform. This is made of a closely woven mesh, without sticky silk, and embedded in a dense mesh of upper and lower scaffolding silk lines. Araneids use non-sticky silk to wrap their prey, usually a flying insect; they rotate it with the third pair of legs and use the fourth pair to draw a band of silk from a brush of aciniform gland spigots on their spinnerets.

Some araneids, such as *Argiope*, apply dense swathes of silk to their webs. These decorations, called stabilimenta (singular, stabilimentum), have evolved independently multiple times in orb-weavers, both ecribellate and cribellate. They may thus serve different functions in different groups.

RIGHT | A female *Araneus diadematus*, in a typical position at the hub of its orb web.

GENUS
Araneus

DISTRIBUTION
Cosmopolitan, except for Australia and New Zealand

HABITAT
Gardens, orchards, and other areas of vegetation; hides in leaves or under bark

CHARACTERISTICS

- Vertical orb webs with a viscid, sticky silk spiral
- Posterior median eyes with a modified, canoe-shaped tapetum (light-reflecting structure)
- Male palp rotated, showing most sclerites to the side of the cymbium, with paracymbium fused to the cymbium, a radix, and a median apophysis
- Epigynum often extended into an elongated projection called a scape

ARANEIDAE: *NEPHILA*

GOLDEN SILK ORB-WEAVERS

Nephila species are easily recognized by the large size of the adult females and by their large, asymmetrical orb webs, constructed with golden silk. The webs can reach a yard (0.9 m) in diameter and have very long silken support lines. In some species, the females aggregate and their webs become interconnected, covering a large area—this makes the females even more conspicuous to males. Around 20 species of *Nephila* have been described worldwide, from a wide diversity of habitats in tropical and subtropical regions.

Nephila are robust spiders with long legs, and the females are often quite colorful. *Nephila* webs, as well as those of their close relatives—such as *Nephilengys* and *Herennia*—are characterized by the retention of the non-sticky, temporary spiral in the finished web. In contrast,

LEFT | A female *Nephila plumipes* with a much smaller male in her web, from Queensland, Australia. Adult female nephilines, especially those in the genus *Nephila*, are gigantic relative to the males.

GENUS
Nephila

DISTRIBUTION
Tropical and subtropical regions

HABITAT
Webs on vegetation in a wide range of habitats, from rainforests and gardens to shrublands and mangroves

CHARACTERISTICS
- Largest orb-weavers (females reaching c. 1.5 inches/40 mm body length
- Extreme sexual size dimorphism, with large females and small males
- Striated patch on chelicerae

this temporary spiral is removed in the webs of other araneids as the sticky spiral is being laid. *Nephila* spiders spend most of their time at the hub of their large webs, waiting for prey to be intercepted during day and night.

The body of females of the largest species of *Nephila* reaches around 1.5 inches (40 mm) in length, whereas the adult males are very small. In the species with the most extreme sexual size dimorphism, the female body can be more than 11 times the size of that of the male and about 100 times heavier. Although *Nephila* males are often regarded as dwarfs, this dramatic difference in body size between the sexes has evolved through increasing female size and not by male dwarfism. This is hardly surprising since *Nephila* females are the largest araneids and nephiline males are not particularly small relative to the size of other araneid species.

Upon reaching sexual maturity, *Nephila* males invade immature female webs, where it is common to see several suitors waiting for the female to mature. Females often mate with more than one male. Nephiline sexual biology involves a wide array of strategies, including mate guarding, sexual cannibalism, genital mutilation, mating plugs, and male castration. Small kleptoparasitic spiders of the theridiid genus *Argyrodes* can be often found in the webs of some *Nephila* species. Here they take advantage of small prey items that are captured in the web but ignored by the host.

Nephila silk is so strong that the webs can ensnare not only very large insects, but even small birds and bats. In some regions, such as Papua New Guinea, *Nephila* webs are used by local communities as nets to catch small fish. The silk has also been harvested to produce textiles, including a handwoven and embroidered golden cape recently made using silk from 1.2 million female *Nephila madagascariensis*; it took three years to craft. Biomedical researchers are actively studying *Nephila* silk as a biomaterial for tissue engineering, as it may be useful in helping the regeneration of human nerves and cartilage. Spider silk can support the attachment and proliferation of a variety of cell types, and could thus revolutionize medical applications in the future.

RIGHT | A female *Nephila pilipes* from northern Queensland, Australia. The adult males are only a fraction of the size of the gigantic females.

ARANEIDAE: *MASTOPHORA*

BOLAS SPIDERS

Mastophora includes about 50 species from North, Central, and South America. Females can reach 0.75 inches (19 mm) in length, while the adult males are smaller than 0.067 inches (1.7 mm) and reach maturity after two instars (or fewer).

Although their closest relatives spin orb webs, bolas spiders owe their common name to their remarkable foraging strategy. This uses a single silk line reminiscent of the bolas used by the gauchos of Argentina and Uruguay to catch cattle.

GENUS
Mastophora

DISTRIBUTION
Mastophora species range from southern Canada to Argentina; other bolas spiders are known from most parts of the world, except temperate Eurasia

HABITAT
Branches and leaves of trees, bushes in orchards and gardens, or along fences

CHARACTERISTICS

- Cephalothorax often with dorsal protuberances, frequently paired
- Female abdomen large, wider than long; on occasion has dorsal humps or lobes
- Extreme sexual size dimorphism: adult females much larger than males
- Prey specialization by means of aggressive chemical mimicry

RIGHT | A female Malagasy bolas spider, *Exechocentrus lancearius*, holding the silk line with its glue droplets to capture prey.

LEFT | A female *Mastophora cornigera*. Adult male bolas spiders are only a fraction of the size of females.

Late-instar and adult females specialize in hunting male moths at nighttime. As *Mastophora* spiders hang from a horizontal silken line, they swing a silk line, or "bola," made of coiled silk and glue, with a sticky globule at its end. Once this globule strikes a moth, it rarely escapes. Bolas spiders can attract their prey, in complete darkness, because they emit chemical components that mimic the sex pheromones of the moth species. *Mastophora* species use the first pair of legs to manipulate the bolas line, but their relatives on other continents (*Cladomelea*, *Ordgarius*, and *Exechocentrus*) manipulate it with the second pair of legs; they may also have more than one sticky globule on their foraging lines. The bolas-swinging behavior differs between genera, and some species whirl the bolas line.

Bolas spiders are uncommon and difficult to find. During the daytime *Mastophora* species cryptically rest exposed, on leaves and twigs, and the adult females of several species mimic bird droppings or leaf buds. Adult males and young of both sexes attract psychodid flies, which they capture without the use of silk from leaf margins, using their first two pairs of legs to grab the prey. Females lay a single egg sac at a time, but multiple egg sacs are clustered over a period and suspended by strong threads on branches.

PHYXELIDIDAE: *VIDOLE*

LACE-WEB SPIDERS

LEFT | A subadult female *Vidole capensis* from Cape Town, South Africa. This is the most common cribellate spider in many habitats along the southern coasts of South Africa, and is often found under logs or tree bark.

Phyxelididae is a family from the Old World, most diverse in East Africa and Madagascar. *Vidole* has five species, all endemic to South Africa. They build cribellate webs in holes in trees or logs, leading to a tubular retreat. They have a stridulatory mechanism involving their chelicerae and palps, probably used for communication during courtship.

An interesting phyxelidid species is *Ambohima andrefana*, endemic to Madagascar. The copulatory organs of the male have a very long embolus, matched by long, coiled copulatory ducts in the female. The mating ritual is peculiar. First, the male embraces the female with his long anterior legs in a characteristic way—the metatarsi I and II of males have a special clasping depression in front of a large spine. While the male is clasping the female palpal femur with his metatarsus II, she gently bites his modified metatarsus I.

GENUS
Vidole

DISTRIBUTION
South Africa

HABITAT
Forests, beneath tree bark, or under stones or logs

CHARACTERISTICS
- Cribellate, the legs with three claws
- Thorn-like setae on palpal femora
- Male metatarsi I and II with clasping structures

MALAGASY CRACK-LEG SPIDERS

Udubidae is a small family restricted to West Africa, Madagascar, and Sri Lanka. Udubids are closely related to zoropsids, a group in which the calamistrum is also an oval patch of setae. The genus *Uduba* is endemic to Madagascar. There are only three described species, but specialists know of many more that remain to be described. The spiders build a silken tube extending out of a burrow in the forest floor, from which some cribellate lines radiate. The tube is lined with mosses and is very hard to spot. Some *Uduba* species do not have a cribellum and actively hunt their prey instead.

BELOW | A male *Uduba*, a genus of spiders that live on the forest floor in Madagascar. Udubids are atypical spiders since they possess both a cribellum, a structure to spin adhesive webs, and scopulae on the legs, which are used to grasp and subdue prey.

GENUS
Uduba

DISTRIBUTION
Madagascar

HABITAT
Forest floor

CHARACTERISTICS

- Males have a "tibial crack," a weak line near the articulation with the patella
- Cribellate species with an oval calamistrum
- Tarsi with three claws and scopulae

TITANOECIDAE: *TITANOECA*

ROCK WEAVERS

GENUS
Titanoeca

DISTRIBUTION
Holarctic

HABITAT
Open and sunny places, under stones or between rocks

CHARACTERISTICS
- Cribellate spiders with a divided cribellum and long, uniseriate calamistrum
- Males with complex palpal tibia, including pro- and retrolateral apophyses
- Tarsi without trichobothria (in contrast to members of the RTA-clade)

Titanoecidae is a rather small family with only five genera and 53 species worldwide. Most of them are distributed in the northern hemisphere, but a few occur in the Oriental and neotropical regions. They all build irregular cribellate capture webs, mostly under stones, or space webs between rocks—as in *Anuvinda escheri*, which occurs in South and Southeast Asia. The genus *Titanoeca* consists of roughly 30 species occurring in the Holarctic. Their coloration is more or less dark brown to black; the cephalothorax and parts of the legs are usually lighter. Many species have a characteristic pattern of distinct white patches on the abdomen; this varies between species, and sometimes also within an individual species. Males generally have more or less distinct patches; the females' abdomen is often uniformly black. Favored habitats are predominantly open and sunny.

One special case is the species *Goeldia zyngierae*, described from a cave in Brazil. It belongs to the only genus with representatives in the neotropical region. These spiders build webs on cave walls or on the ground. The webs, up to 16 feet (5 m) in length, can be at least partially communal, with two to 30 spiders of different stages and sexes moving freely about in the interconnected web. When the web is disturbed, the spiders approach the disturbed area together, suggesting that this species shows at least some social behavior.

LEFT | This male *Titanoeca* shows how spider species receive their names; the four white patches on the abdomen suggested to Carl Wilhelm Hahn in 1833 the species name *quadriguttata*, meaning "four times spotted."

RIGHT | A female *Titanoeca tristis*, showing the typical uniform coloration of the sex. The phenomenon of males and females having different patterns or colors is called sexual color dimorphism.

THE "RTA-CLADE"

The "RTA-clade" contains some 40 families of entelegyne araneomorphs, which combine to represent almost as many species as all other spider families together. The only character that unites these spiders is the presence of a retrolateral tibial apophysis (RTA), a projection on the tibia of the male palp. It is not certain if this single feature renders these spiders so successful, or why it may do so, but it certainly highlights the importance of the genitalia and their structure. This group appears to contain many efficient morphological forms, sometimes called "templates," with many similar species, differing only by the shape of the palp and epigynum.

LEFT | A crab spider in one of its favorite habitats, on a flower, waiting for insect prey to arrive.

SIMPLE AND COMPLEX PALPS AND EPIGYNA

Males of spiders belonging to this group of families are easy to recognize by the presence of a retrolateral apophysis on the tip of the palpal tibia. The sketches below illustrate the male and female genitalia of two species from the genus *Hortipes*, showing that, like the palpal bulb and the female epigynum, the retrolateral tibial apophysis can vary from simple to complex within the members of a single genus.

SIMPLE PALPS AND EPIGYNA
(*Hortipes castor*)

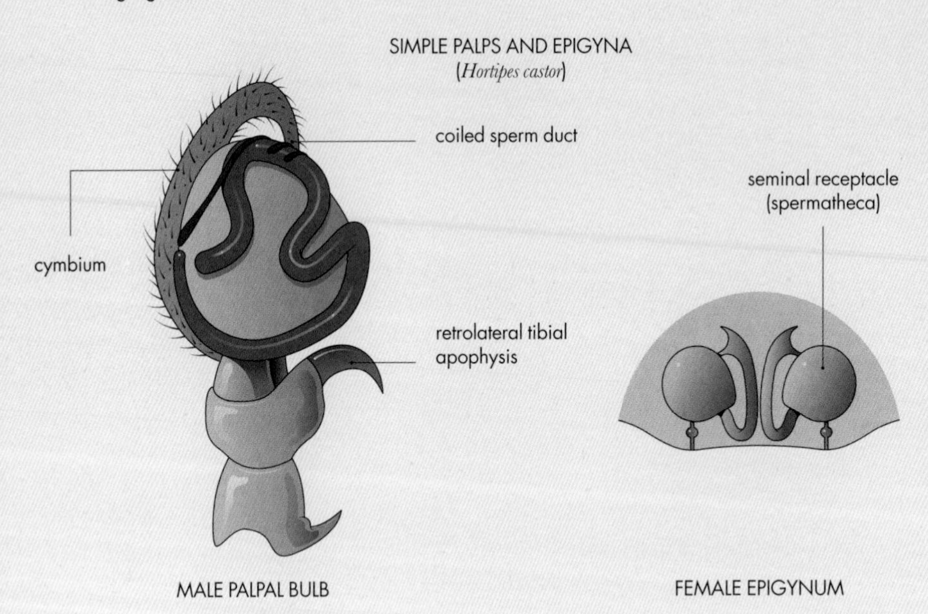

COMPLEX PALPS AND EPIGYNA
(*Hortipes sceptrum*)

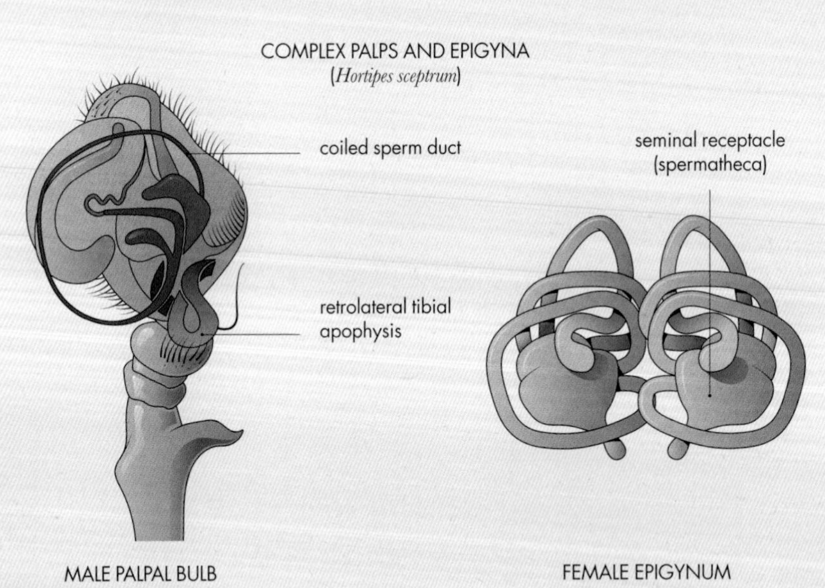

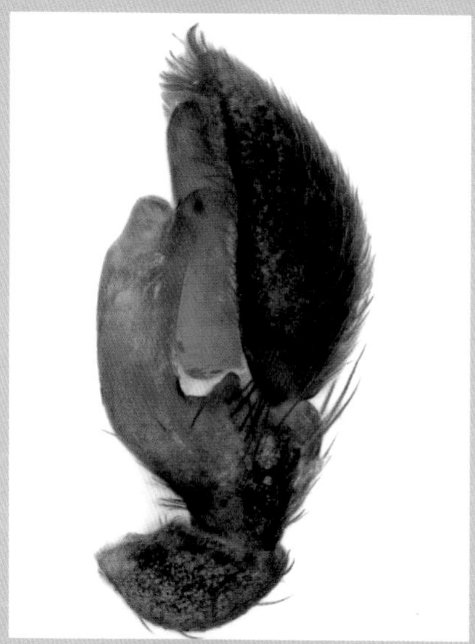

Such is the case for genera such as *Hortipes*, *Mallinella*, and many others treated in this volume. Their genitalia show an enormous range of complexity, with elongation of the embolus as a common tendency. The RTA appears to play a crucial role in arresting the rotation of the bulb during copulation, and thus helps to lock the embolus in a particular position that fits the entrance to the epigynum. This is clearly a system that facilitates radiation within these templates, based on differences among the many species. The rationale behind the necessity for these differences is subject to much debate.

ABOVE | Retrolateral views of the male palps of members of four different families belonging to the RTA-clade, showing a wide range of forms of the RTA (upper left, *Clubiona comta*, Clubionidae; upper right, *Cryphoeca silvicola*, Cybaeidae; lower left, *Amaurobius fenestralis*, Amaurobiidae; lower right, *Phrurolithus festivus*, Phrurolithidae).

DICTYNIDAE: *ARGYRONETA*

DIVING BELL SPIDER

Argyroneta aquatica is the only spider known to live in fresh water. Both males and females construct a silken "diving bell" attached to underwater plants. They fill it with air, brought down from the surface in large bubbles and held with hairs on the abdomen and rear legs. The air is trapped by surface tension between the silk fibers, and the bell is open at the bottom. The diving bell functions as a physical gill, allowing dissolved oxygen to enter the bell from the water, while the carbon dioxide produced by the spider dissolves in the water.

Interestingly, although the water can supply sufficient oxygen to support the animal's resting physiology, even in warm, stagnant ponds, nitrogen also has an outward gradient into the water. The spider must thus replenish the diving bell with air from the surface to keep it from shrinking. During warm summer months, the spiders may replenish the bell two or three times an hour. However, they can remain in the bell continuously during the winter, effectively hibernating.

Argyroneta aquatica is also one of the few spiders in which males are larger than females, weighing about 50 percent more. The larger males have more mobility underwater; after their last molt, they charge their palps with

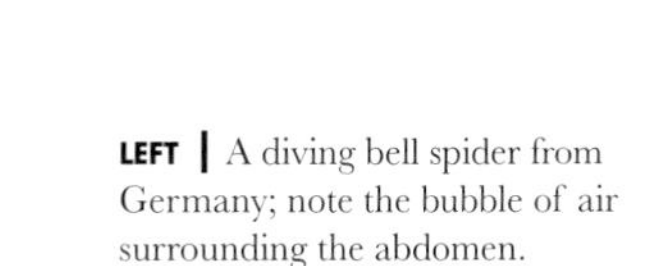

LEFT | A diving bell spider from Germany; note the bubble of air surrounding the abdomen.

RIGHT | A diving bell spider, seen inside its underwater bell.

GENUS
Argyroneta

DISTRIBUTION
Europe and parts of Asia, from Britain to Japan

HABITAT
Fresh water with a low current and underwater plants

CHARACTERISTICS
- Lives almost entirely underwater
- Has hydrophobic hairs that trap an air bubble when the animal submerges
- Both sexes construct—and live in—an underwater diving bell

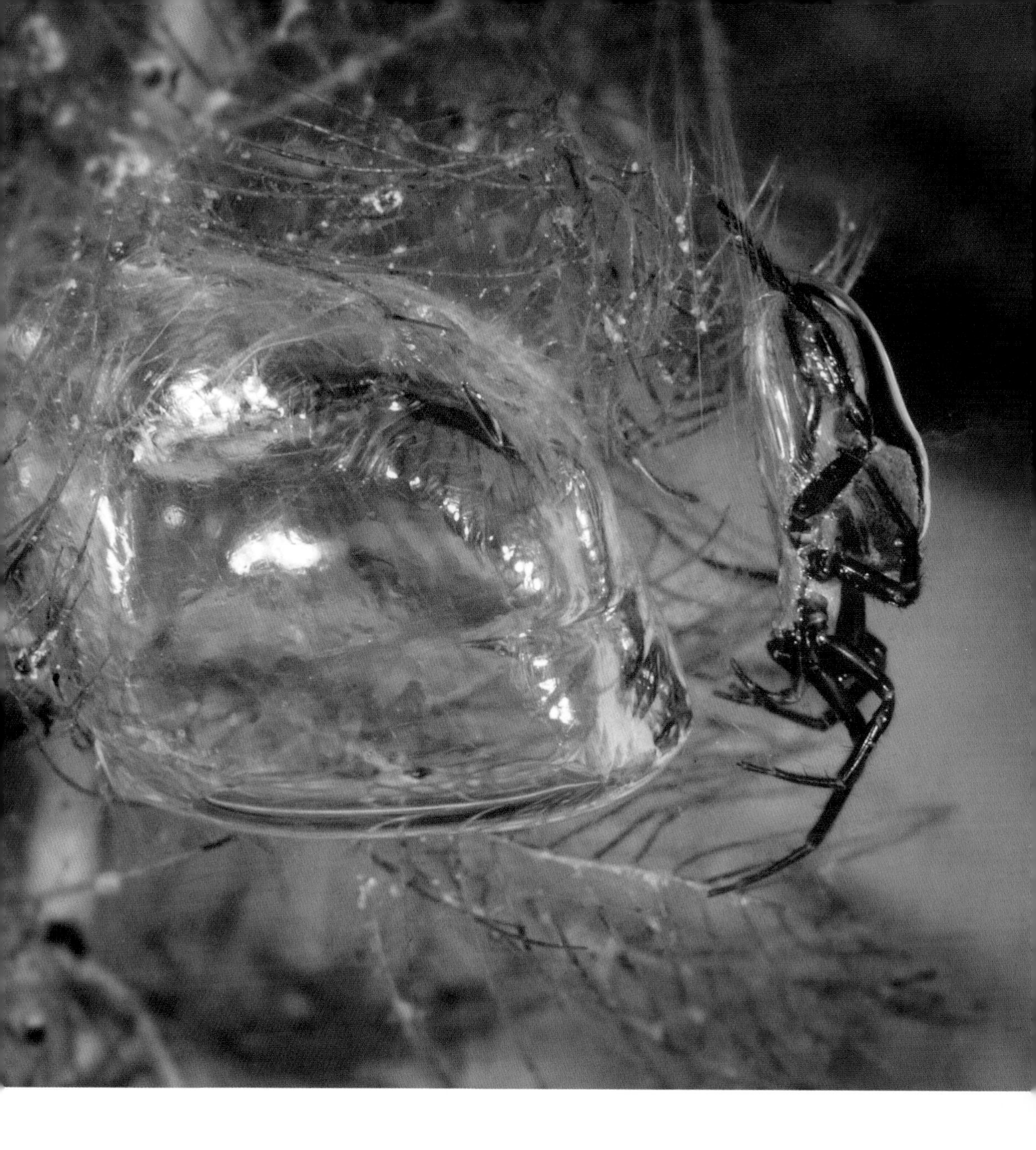

sperm and leave their diving bell in search of a female's. Mating takes place inside the female's bell, after which the female constructs an egg sac in the upper part of the bell. Females that have mated only once can produce as many as six sacs containing up to 100 viable eggs per egg sac/clutch. During the three–four weeks before the spiderlings hatch, the mother seldom leaves the bell and narrows its entrance.

Males are more active hunters, and the spiders are important predators of aquatic crustaceans and insects (including mosquitoes), which occur in waters with low pH and little dissolved oxygen (where few fish can survive). The many other dictynids, found worldwide, typically build tangled webs in low vegetation.

ZODARIIDAE: *MALLINELLA*

FOREST ANT-EATERS

Spiders of the genus *Mallinella* belong to the Zodariidae, one of the larger spider families, and are called burrowing spiders or ant-eating spiders. *Mallinella* are among the most abundant hunters in the leaf-litter layer of tropical rainforests within their huge distribution area. So far, almost 200 species have been attributed to the genus, but many more are awaiting description, mainly from the Afrotropical realm.

These spiders are predominantly active at night and feed almost exclusively on ants and termites. No fewer than six species have been found to cohabit in some forests; it is therefore assumed that many have a specialized diet, feeding on particular species of those insect groups. In order to avoid contact with these aggressive prey species during the day, they rest in a small, silk-lined hole dug out of the deeper litter layer. The presence of a row of tiny spines in front of the tracheal spiracle has been linked to this behavior, as the spines may safeguard this crucial orifice from becoming clogged with soil debris.

LEFT | The majority of the numerous *Mallinella* species have a medium- to chestnut-brown carapace, but some—like this male from Mt. Nimba in West Africa—have a reddish tinge.

RIGHT | *Hermippus* is most particular about the type of litter it lives on; this species has a vast distribution in miombo woodland but avoids litter of adjacent montane forest.

GENUS
Mallinella

DISTRIBUTION
Old World forests

HABITAT
Tropical rainforests, in leaf litter; some species prefer mountain forests/woodlands with a well-defined dry season

CHARACTERISTICS

- Dome-shaped carapace
- Row of small spines in front of tracheal spiracle
- Constructs a retreat in leaf litter

A related genus, *Hermippus*, contains a dozen species. It has a similarly wide distribution in tropical areas of southern and eastern Africa, India, and Sri Lanka. These spiders also live in the leaf litter, but occupy habitats different from those frequented by *Mallinella*: they prefer woodlands with a well-defined dry season. The two genera seem never to occur together, even though they may live near one another, where woodland and mountain forest alternate.

In the field, *Hermippus* are very conspicuous, due to the contrast between their white trochanters and their high, domed, dark brown carapace. Unlike other zodariids, which have three tarsal claws, *Hermipppus* have only two, accompanied by claw tufts. They also have a strong, protruding chilum, the sclerite situated between the chelicerae and the front margin of the carapace, and peg teeth on the cheliceral margin—these are modified spines, very different from the ordinary cheliceral teeth, which are sharp outgrowths of the integument. Whether *Hermippus* are as species-specific as *Mallinella* in their feeding habits is unknown.

ZODARIIDAE: *CRYPTOTHELE*

WALKING MUD SPIDERS

Cryptothele is a small genus, with only six known species, and is one of the few zodariids other than *Mallinella* to inhabit leaf litter. These unusual spiders have a restricted distribution in forests of the Far East but, surprisingly, also occur on the isolated granitic islands of the Seychelles. Very few spiders apply the strategy that *Cryptothele* use to remain unnoticed. Their bodies are covered with sticky hairs, to which small leaf and soil particles adhere. This makes the animal virtually invisible to the inexperienced eye, and they are therefore called walking mud spiders. Vital parts of the abdomen, including the spinnerets and book lungs, do not have the adhesive hairs. Instead, thick setae protect these parts from contact with the soil debris that otherwise coats the animals.

Like many other zodariids, *Cryptothele* seem to be specialized feeders on termites. There are other unusual aspects of their morphology, including a flat carapace with a deep pit behind the eyes, a rare abdominal shape, and paired tarsal claws with few teeth. Because of such features and their bizarre appearance, the genus was for a long time placed in its own family, until its relationships with other

RIGHT | Few spiders use the strategy of the walking mud spiders to stay unnoticed: they cover themselves with soil and leaf litter debris so that they merge with their surroundings.

GENUS
Cryptothele

DISTRIBUTION
Australia, large parts of Indonesia, Fiji, Samoa, New Caledonia, Seychelles, Sri Lanka

HABITAT
Forests, in leaf litter; rocky granite islands of the Seychelles

CHARACTERISTICS
- Flat carapace with a deep pit behind the eyes
- Eye arrangement in three rows: 2–2–4
- Body covered with clusters of thick, sticky hairs
- Unusual abdomen shape

zodariids were determined. Their characteristic but rare eye pattern, in three rows (2–2–4), clearly links them to members of a particular subfamily of zodariids that share this unusual arrangement.

Species of *Psammoduon*, which inhabit the Namib Desert and other sandy areas of southern Africa, have an even more surprising way to disappear—they literally dive into the sand. To do so, they flip on their back and wriggle their front legs, which are provided with long, supple hairs, to drive themselves down into the sand. The posterior leg pair provides the thrust to push the spiders deeper, until they are completely covered. This behavior has earned them the name back-flip spiders (although "sand-swimmers" might be just as appropriate, since they are known to be able to move under the sand). They feed on the sand surface, mainly capturing larvae of tenebrionid beetles, but plunge into the substrate as soon as the temperature exceeds 77°F (25°C). This behavior helps to protect the *Psammoduon* not just from high temperatures, but also from huntsman spiders of the genus *Leucorchestris*—the dominant invertebrate predators in these sandy habitats.

ABOVE | Walking mud spiders of the genus *Cryptothele* have a remarkable distribution, far away from their nearest relatives. Together with their unusual morphology, this explains why they have only recently been recognized as zodariids.

ZODARIIDAE: *ZODARION*

IGLOO ANT-EATERS

The many species of *Zodarion*—a large, Palearctic genus of compulsory ant-eaters—have yet another way to hide. They build an igloo-shaped, silk-lined retreat from tiny pebbles or sand grains, then attach it to the underside of a stone or log. At dawn, the spiders leave their retreat in search of ants, which they most often attack near the entrance to their nest. The cheliceral fangs of these spiders are transformed into short, dagger-like syringes that inject venom in the prey's legs during a rapid attack. The spider withdraws until the ant has collapsed, then returns to start feeding on it.

These poor-sighted ant-eaters all have tiny glands on the extremity of the femora. It is possible that the product emanating from these glands is a warning system, preventing attacks by fellow ant-eating zodariids that might mistake them for prey.

RIGHT | *Zodarion* is a genus of compulsory ant-eaters, attacking prey that is sometimes many times their size.

GENUS
Zodarion

DISTRIBUTION
Palearctic, from west Europe to Mongolia, including the Canary Islands

HABITAT
Dry areas with sparse vegetation

CHARACTERISTICS
- Large anterior median eyes
- Legs densely covered with flattened hairs
- Femora with a distal gland

SAND SPIDERS

Homalonychus is the only genus in one of the smallest families, and is known from only two species. They live in deserts and arid environments in the western part of North America, on both sides of the Mexico–USA border. In resting position, these spiders are camouflaged by sand grains lodged between the stiff setae on their backs. They pick up the grains by flipping over quickly in a shallow cavity, made by scratching in the sand.

The spiders' habit of partly burrowing themselves in the soft, sandy substrate on which they live explains why they have been considered relatives of the zodariidae. However, the details of their morphology suggest otherwise. They have smooth tarsal claws, claw tufts, and accessory claw setae—a combination found nowhere else among spiders. Accessory claws are usually associated with web construction, but these spiders do not even make draglines, another rare exception. Silk is used only by the female for egg sac construction and, remarkably, by the male to wrap the female during copulation and to make sperm webs.

RIGHT | Sand spiders of the genus *Homalonychus* share the habit of *Cryptothele*, covering themselves with material from their surroundings. Since they live in sandy deserts, they typically use sand grains to do so.

GENUS
Homalonychus

DISTRIBUTION
Northwestern Mexico, southwestern USA

HABITAT
Sandy deserts and arid environments

CHARACTERISTICS
- Tarsi have a pair of smooth claws, and a pair of accessory claw setae
- Sternum broad, with a posterior protrusion
- Do not make draglines or webs

STIPHIDIIDAE: *STIPHIDION*

SOMBRERO-WEB SPIDERS

Most spider webs have straightforward shapes, usually forming some sort of catching sheet (circular or not); some lead back to a tough, thimble-shaped silken retreat. The web of *Stiphidion* deeply defies that convention. Imagine four short ice-cream cones set in a square pattern, with a large sheet across the top and the spider hunting from the space formed at the bottom by the four cones. That is the web of *Stiphidion*, a spider happiest in the cooler forests of Australia and New Zealand.

The optimal place for such a web is between two roughly parallel surfaces—something rarely seen in nature except between boulders or under dislodging bark. Hence, although the spider greatly appreciates such gaps formed around our huts and seats, without those surfaces the web it makes is somewhat

GENUS
Stiphidion

DISTRIBUTION
Eastern and southern Australia, New Zealand

HABITAT
Cooler forests; webs hang among stones and tree bark, occasionally on man-made structures

CHARACTERISTICS
- Three tarsal claws
- Eyes in two recurved rows
- Ornate striped pattern
- Raised head area
- Long legs
- Spinnerets extend from the rear
- Irregular web

LEFT | Sombrero spider (*Stiphidion facetum*). The rusty red-brown body, flanked by a series of white lines, is indicative of this group.

ABOVE | A specimen of *Stiphidion* on its web, which has no well-defined pattern but overall is composed of a single curving sheet.

cramped. The spider hunts from a number of locations on the web: in the central cone, on the web, and on the walls. Only four species are included in the genus, but almost 20 other genera from Australia, New Guinea, and New Zealand are currently assigned to the family.

Stiphidion species have an eye pattern otherwise seen only in free-ranging, hunting spiders. Almost all other members of the family have a conservative eye arrangement, with two rows, each of four similar eyes, that are curved like a lens. In *Stiphidion*, the eyes of the back row are larger than the front eyes and the back row is curved back, in a typical hunting spider formation.

Similarly, no other spider makes a web at all like that of *Stiphidion*. The other genera in its family all make a rudimentary framework or superstructure from which a simple sheet web is suspended; it then narrows to a funnel (either above or below, depending on the genus). The deepest point of the funnel is protected in a soil crevice or tree bark. The spider rests in the funnel and constructs its egg sac there, lunging out when prey strikes the superstructure.

AGELENIDAE: *ERATIGENA*

FUNNEL-WEB HOUSE SPIDERS

BELOW | A female *Eratigena atrica*, guarding two egg sacs, one (right) freshly laid, the other (left) already camouflaged with litter.

These impressive spiders are well known in Europe and North America as funnel-web spiders, living—in addition to their natural habitats—in and around buildings. They are easily noticed, especially when wandering spiders are trapped in sinks, shower trays, and bathtubs. However, they are harmless, biting very rarely and usually causing no symptoms in humans.

The so-called Hobo Spider (*Eratigena agrestis*) has sometimes been thought to cause necrosis through its bite, but there is little evidence to support this claim.

Eratigena spiders build large sheet webs with the aid of their long spinnerets. The webs have a funnel at one side leading to a tube-like retreat. Here, the spider waits until prey runs onto the web and attracts its attention by causing vibrations, transmitted by the web into the retreat. The spider immediately sprints to the prey and attacks it with a bite. It may wait for the injected venom to paralyze the prey before pulling it into the retreat. When running across the web, the spider

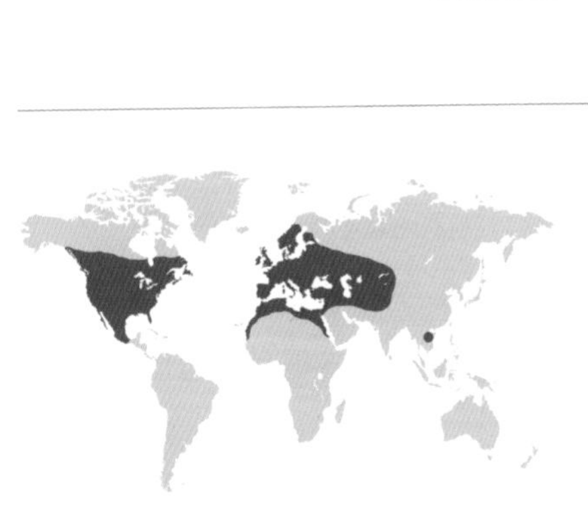

GENUS
Eratigena

DISTRIBUTION
Holarctic (Canada, USA, Mexico, Europe, North Africa, central Asia, Laos)

HABITAT
Dry places such as hollow trees, and caves; often close to, or in, buildings

CHARACTERISTICS
- Builds funnel webs (often in and around houses)
- Spinnerets with elongate distal articles
- Camouflages egg sacs with soil and plant particles

avoids becoming entangled in the threads by setting its tarsi almost vertically on the web, so that only its claws and some distal spines come into contact with the silken sheet. This method cannot be used by the prey, so the spider is always much faster. *Eratigena* spiders are probably the fastest runners among web-building spiders, capable of covering more than half a yard (0.5 m) per second.

When a male enters a female's web, both sexes may live together for a while and mate several times, until the female lays her egg sac. First, the spider produces a vertical silken plate on which the eggs are laid. Then it weaves silk to embed the eggs in a layer that helps to insulate them against winter cold. The external layer is stronger again, and covered with particles such as sand, leaf litter, and prey debris; these camouflage the egg sac from predators and parasites. In the next season, the hatching juveniles build their first webs within the mother's large web, before leaving to seek their own territories.

LEFT | A female *Eratigena atrica* sitting on a wall. These harmless spiders often inhabit houses and their surroundings.

DESIDAE: *DESIS*

INTERTIDAL SPIDERS

From the shores of southern and eastern Africa and the Seychelles to Japan and Samoa, from the tropics of Queensland to temperate parts of Tasmania and New Zealand, and on the Galápagos Islands, on deeply rooted kelp and coral reefs, these small spiders have almost returned to the sea, making the intertidal zone their habitat of preference. Their very long chelicerae and fangs no doubt help them catch and kill potentially dangerous prey, including small crabs and isopods. These spiders have no special respiratory modifications (such as gills), but they do have very long, curled hairs on their legs.

Some *Desis* species, such as *Desis bobmarleyi*, found near Cairns in northern Australia, live on fringing reefs. These are exposed to the air only at the four or five maximally low tides, dropping at least to 1 foot (0.3 m), that occur each year;

LEFT | A *Desis bobmarleyi* female, showing long, curled hairs on its legs that may help hold air.

RIGHT | *Badumna* species build lacy, funnel-shaped webs.

GENUS
Desis

DISTRIBUTION
Shorelines, from the tropics to cold temperate regions

HABITAT
Intertidal areas, from kelp and coral reefs to fringing reefs that are exposed to the air only infrequently

CHARACTERISTICS
- Small intertidal spiders, some submerged for months
- Long, curled hairs on the first two pairs of legs
- Long, slender chelicerae and legs

the species name thus alludes to Bob Marley's song, "High Tide or Low Tide." Like all spiders, it liquefies its prey and needs air in which to breathe and feed. Presumably, turbulence at "normal" low tides generates sufficient air cavities through the coral, where the spiders construct a thin sheet of air-trapping silk.

At least in the lab, *Desis bobmarleyi* hunts at the water–air interface. It lurks upside down in the water with its special, prey-sensing hairs in the water's meniscus, sensing vibrations made by insects on the surface. Many species of *Desis* occur in the intertidal zones of shorelines; one was even reported by puzzled British marine scientists at a depth of 33 feet (10 m) in the Red Sea.

Other spiders that are currently placed in the Desidae family include the Australian *Badumna*, which build conspicuous, lacy, funnel-like webs in deep, protected spaces under thick eucalypt bark. As humans built houses near forests, these dark, bushy-legged spiders began finding similar spaces in the many corners around windows and doors—even on the many gaps on cars and other vehicles. With human movements of infested objects, such as shipping containers, from Australia, the world has become the oyster of these brown house spiders, which can now be found in many other countries, including New Zealand, Japan, the USA, Mexico, Uruguay, and Argentina.

In the closely related Australian genus, *Phryganoporus*, maternal care is evident, as the spiderlings long remain in the female's web, forming large, football-sized clusters on bushes.

AMAUROBIIDAE: *AMAUROBIUS*

MESHWEB-WEAVERS

Spiders of the genus *Amaurobius* are small to medium-sized and generally brown in color. The 60-plus described species live beneath stones and logs, under bark, between rocks, or in crevices in walls. They build their web from a hidden retreat and these are characterized by the special woolly silk produced by the cribellum. This structure is located in front of the spinnerets and releases fine

BELOW | A female *Amaurobius fenestralis*, showing the typical dark pattern on white guanine crystals on the abdomen. Note the characteristic, slightly bluish cribellate silk strands on the right. These are seen outside the retreat and in most cases members of the genus can be identified from their web alone.

RIGHT | *Amaurobius ferox* is another widespread European species that was introduced to North America. It is larger than *A. fenestralis* and prefers humid, shady places to build its web, such as rock crevices, caves, walls, and basements of houses.

GENUS
Amaurobius

DISTRIBUTION
Holarctic

HABITAT
Under stones, logs, and bark, between rocks, in wall crevices

CHARACTERISTICS
- Produce webs with bluish cribellate wool for catching prey
- Calamistrum has a double row of bristles
- Females may be eaten after brood care by spiderlings (matriphagy)

threads with diameters of just 0.01–0.02 μm. The silk is pulled from the cribellar spigots by the calamistrum, a double row of serrated bristles situated on the last leg, resulting in a shimmering whitish to bluish mesh.

The silk contains no glue like that found in the webs of orb-weaving spiders. Rather, its adhesive effect is driven by Van der Waals and hygroscopic forces. Recently, it was discovered that cribellate catching silk interacts with the semifluid wax present on insects' exoskeletons. The silken mesh absorbs this liquid, binding with the chitinous armor of the prey and decreasing their likelihood of pulling free from the web.

A special behavior of *Amaurobius* spiders is seen when females are eaten by their offspring (matriphagy). Experiments have shown that spiderlings that feed on their mother are larger, heavier, and more successful; they showed a more extended social period than those in a control group without matriphagy, even though the individuals in the control group were also well fed. Thus, a greater total reproductive success seems to be the evolutionary advantage of the matriphagous lineages.

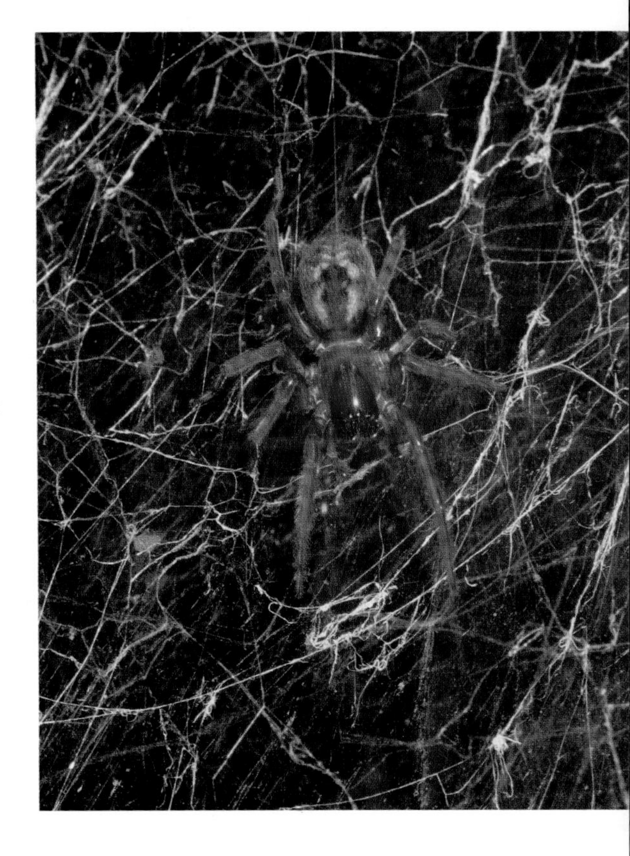

ABOVE | *Amaurobius fenestralis* in its web, showing the characteristic appearance of the cribellate silk.

ZOROPSIDAE: *ZOROPSIS*

FALSE WOLF SPIDERS

False wolf spiders have earned this name because their looks and behavior have several elements in common with lycosids, the typical wolf spiders. They belong to a relatively small family, Zoropsidae, with fewer than 200 species. Although *Zoropsis* species have a well-developed cribellum, their silk does not serve for the construction of a catching web. Rather, they use their tough cribellate silk primarily to build a retreat and to protect the egg sacs. Prey is subdued in a manner typical of wolf spiders: the legs, provided with rows of erectile spines on the tibiae, surround the insect before it is bitten, a posture that has been called "the leg basket." False wolf spiders have dense scopulae (clusters of modified hairs) that enhance the spider's grip on struggling prey. They also possess claw tufts, which enable a sneaky approach to the prey from vertical surfaces. This contrasts with the attack of real wolf spiders, which pounce with great speed on nearby prey, most often from a horizontal substrate.

Zoropsis spinimana is one of the species that has greatly expanded its originally Mediterranean range. Just a few years ago, it was rarely observed in buildings in western Europe. Today, it has become part of the established fauna of that region and is frequently found in gardens.

LEFT | *Zoropsis spinimana* has recently expanded its range considerably to the north and has become common in western Europe and even in a part of California, USA.

RIGHT | The impressive but innocuous *Zoropsis spinimana* has unnecessarily raised concern because it often seeks shelter in houses in its newly conquered area.

GENUS
Zoropsis

DISTRIBUTION
Southern Palearctic; introduced to the USA (central California)

HABITAT
Originally, some species had a Mediterranean habitat; now much expanded into western Europe in the wild, on buildings, and in gardens

CHARACTERISTICS
- Robust cribellates
- Two tarsal claws, claw tufts, and scopulae
- Spins a tough cribellate retreat under stones

MITURGIDAE: *MITURGA*

RACING STRIPE SPIDERS

Presently including only 17 species (with more than 100 still to be described), this genus of striped spiders has diversified strongly in Australia's outback. Many look very similar to the one illustrated here; as with many spider groups, they differ only in the detailed structure of the male and female sexual organs. The medium-sized spiders run at night; during the day, they rest in nests of dense white silk either in the ground, under bark, or in swamps. The extent of the web formed in drying cracks in the earth in Australia is reminiscent of some tarantula burrows, but on excavation proves to be the den of *Miturga*. In one case, the burrow was filled with the dried carcasses of several introduced Cane Toads (*Rhinella marina*), a scourge on the landscape.

LEFT | A female *Miturga lineata*—a widespread predator of Cane Toads (*Rhinella marina*).

RIGHT | Long stripes on the carapace and abdomen are characteristic of most *Miturga* species.

GENUS
Miturga

DISTRIBUTION
Australia

HABITAT
The Outback regions, nesting on the ground, under bark, or in swamps; webs span cracks in dry earth

CHARACTERISTICS
- Two tarsal claws
- Lack claw tufts
- Long stripes down body
- Male palpal tibia with two locking prongs, sandwiching soft cuticle

Miturgids, like the members of a number of large spider families (Sparassidae, Salticidae, Corinnidae, Gnaphosidae, Zodariidae), are "attracted" to the vibrations caused by a poorly tuned, idling diesel engine, such as that of a tractor. Newer 4x4 engines, being too well balanced and with very smooth idling, tend to have little effect. The attraction operates only during the day, and on dry soils. These vibrations seriously aggravate the spiders, which often rise to a defensive pose with their head raised, their chelicerae spread, and their fangs opened; they will occasionally bite non-offending bystanders. Being large, *Miturga* and members of some related genera always impress the uninitiated when they are seen 20 yards (18 m) away, "thundering" across the leaves toward the vibration source. Hundreds of spiders can come to a vehicle within 30 minutes. In one survey, more than 100 spider species of diverse families were taken by this vibration method in a number of locations. On some occasions, however, the vibrations merely cause the spiders to move up and down through the layers of leaves, thus making them more evident.

Miturga has relatives around the world, but the male palp of one Australian species looks eerily similar to that of a species of a different genus that occurs from Israel to Japan. However, those spiders have quite different eyes.

CHEIRACANTHIIDAE: *CHEIRACANTHIUM*

YELLOW SAC SPIDERS

With its 200-plus species, the genus *Cheiracanthium* contains about two-thirds of the total known diversity of the family Cheiracanthiidae (12 genera, 351 species). The vernacular name of the group points to the retreat that sac spiders build to hide in during the daytime. Usually this is woven between leaves or stalks of low vegetation, and it is often built newly every morning. Thus the characteristic retreats can be observed either occupied or empty.

At night, the spiders roam through vegetation in search of prey. Especially on cold nights, when insects can be too cold to fly or jump, some *Cheiracanthium* can subdue animals much larger than themselves. Their relatively strong venom helps them to immobilize the prey. In a few cases, the venom also has an effect on humans. Although bites of most species produce only mild and fleeting symptoms, those of *C. punctorium* and a few other species can cause pain lasting a day or more, nausea, and muscle twitches. Most bites happen when a female is disturbed while guarding her egg sac in a retreat. When bothered, the female will fully open her fangs, producing an impressive warning display.

LEFT | A female *Cheiracanthium punctorium* on top of her retreat, in which she hides during the day. At night, these spiders roam through the jungle of stalks and leaves to hunt for prey.

GENUS
Cheiracanthium

DISTRIBUTION
All continents

HABITAT
In vegetation of open habitats and forests

CHARACTERISTICS
- Males have a characteristic cymbial spur on their palps
- Constructs a retreat near tip of plant leaves and grasses
- Mild to relatively strong venom

An unusual habitat sometimes occupied by *Cheiracanthium* spiders is the air valves of fuel tanks in diesel automobiles. When the retreats are constructed in those valves, the mesh can be tight enough to cause problems with the fuel supply. It is not clear why *Cheiracanthium* seem to favor this hiding place, although the space is the right size for their retreats.

ABOVE | A female *Cheiracanthium punctorium* sitting on the top of an herbal plant. It is one of the rare spider species whose bite causes distinct symptoms in humans.

CYBAEIDAE: *CYBAEUS*

SOFT SPIDERS

Cybaeids are unremarkable brown spiders that live on the soil, in leaf litter, under logs or rocks, and in caves. The family is difficult to distinguish from others in the RTA-clade, and its members have at times been placed in several other families. *Cybaeus* includes more than 150 species in North America, Europe, and Asia, about 40 of which occur in the USA. They construct small, funnel-like webs and are abundant in humid forests and caves.

Other cybaeids common in gardens in North America are several species of *Calymmaria*. They construct a dense, cone-shaped sheet web in cavities, the underside of logs, or wood fences. As in other spiders that build dense sheet webs, they have long posterior spinnerets, probably used to apply many fibers of silk at once to the sheet web, as if using a brush.

Cybaeids have a complicated taxonomy, and the genera are occasionally reassigned to other families, such as Dictynidae, Amaurobiidae, Agelenidae, and Hahniidae.

RIGHT | A female *Cybaeus signifer* feeding on a woodlouse. This nocturnal species is common on the west coast of North America. During the night the spiders wander over tree bark looking for prey.

GENUS
Cybaeus

DISTRIBUTION
Holarctic

HABITAT
Humid forests—in the soil, in leaf litter, under logs or rocks, and in caves

CHARACTERISTICS
- Retrolateral patellar apophysis of male palp has peg setae
- Three claws
- Ecribellate

HAHNIIDAE: *NEOANTISTEA*

COMB-TAILED SPIDERS

Most hahniids are minute spiders that live in small webs constructed over the soil or in leaf litter. They typically have a unique arrangement of spinnerets, with all six placed in a single, transverse row. Many hahniids are adapted to living in the deeper layers of the soil or in caves, and have lost some or all of their eyes.

Neoantistea riparia make small, delicate sheet webs close to the soil in damp areas, usually over small depressions, even animal footprints. For much of the time the web is covered by dew droplets, which apparently help to entangle insect prey.

The genus *Cicurina* is one of the hahniids in which the spinnerets are grouped, rather than set in a row. There are more than hundred species of this genus in the USA, many of them eyeless, endemic to small cave areas, and endangered by development in those areas.

RIGHT | A *Neoantistea* from Texas, USA. Typical hahniids are easily distinguished by having spinnerets spread apart in a transverse line. They are most easily found by sifting leaf litter, or looking for their small sheet webs on the soil.

GENUS
Neoantistea

DISTRIBUTION
Temperate areas of the northern hemisphere

HABITAT
Close to the soil, especially in damp areas, sometimes at deep layers; in leaf litter or caves

CHARACTERISTICS
- Six spinnerets in a transverse row
- Tracheal spiracle large, separated from spinnerets
- Small size

TOXOPIDAE: *TOXOPS*

TASMANIAN SCUTTLE SPIDER

BELOW | A female *Toxops montanus*, hunting among leaves with its front legs directed first toward the sides before extending forward.

Although the genus currently includes only a single species from Tasmania, *Toxops* continues to create problems in determining its evolutionary relationships. Perhaps most indicative of this is that two genera, *Toxopsoides* and *Toxopsiella*, whose similarity led the same author to use "*Toxops*" as the base part of those genus names, are now placed in different families.

Toxops is a small spider, measuring no more than 0.12 inches (3 mm) in length, found in rainforests throughout Tasmania (separate, as yet unnamed, species occur in montane rainforests around an ancient volcanic caldera in southeastern Queensland and in southwestern Western Australia). The spiders were first reported to be living in shrubs with fine leaves (Ericaceae), but were later also found living on the ground—hunting in the litter and walking with trembling front legs. In one *Toxops* pair, the male engaged in a very brief courtship but mated with the immobile female for up to five hours. Only two eggs, about half the size of the adult, are laid.

GENUS
Toxops

DISTRIBUTION
Australia

HABITAT
Rainforests (one species prefers montane forests around ancient volcanic caldera), living on ground among leaf litter

CHARACTERISTICS
- Three claws
- Side eyes of front row are set back beside middle eyes of back row
- Head quite flat

CYCLOCTENIDAE: *CYCLOCTENUS*

SCUTTLING SPIDERS

It was once said that spider classifications work well until the Australian and New Zealand groups are added, and the family Cycloctenidae is a prime example of this. With the exception of one genus and species from Java not seen since 1898 (and thus probably misplaced), the Australasian genera look incredibly different from one another, especially in their eyes.

Cycloctenus includes 17 species, seven from Australia and the rest from New Zealand—although that is probably a very conservative portion of the entire fauna. Instead of having the four eyes of the front row in a more or less straight line, as in most spider groups, the side eyes of the front row are set well back on the slopes of the mound bearing the large middle eyes of the back row. The only other families where this occurs are the similar Toxopidae and the Ctenidae, an apparently unrelated group that occurs worldwide but is most diverse in South America and whose members, very unlike *Cycloctenus*, have dense pads on their tarsi.

Cycloctenus hunt almost exclusively on trees at night, and are more common in the cooler, southern forests. These are spiders with good vision and their legs are set somewhat to the side, as in the huntsman spiders (Sparassidae). When approached, the spiders thus quickly scuttle sideways. Their mottled colors, which include green, provide excellent camouflage against tree bark.

BELOW | In *Cycloctenus*, the anterior lateral are small, and instead of being next to the anterior median eyes are situated much farther back on the carapace, at the sides of the posterior median eyes; the posterior lateral eyes are set even farther back on the carapace.

GENUS
Cycloctenus

DISTRIBUTION
Australia and New Zealand

HABITAT
Cooler southern forests; they hunt on trees

CHARACTERISTICS
- Three claws
- Side eyes of front row set back beside middle eyes of back row
- Head with median saddle or depression

VIRIDASIIDAE: *VIRIDASIUS*

ZEBRA SPIDERS

Viridasiids are a small family consisting of two genera, *Vulsor* and *Viridasius*, endemic to Madagascar and the Comoros. Because of their claw tufts and the eye pattern (in three rows, and equipped with grate-shaped tapeta), they were previously considered tropical wolf spiders (family Ctenidae), as were xenoctenids. Again, genetic evidence finally identified them, suggesting that they are an independent lineage.

Viridasiids are nocturnal hunters, but little else is known of their biology in the wild, although some species are reared and sold in pet shops. Females of *Viridasius* build a pendulous egg sac, suspended from a silk stalk and camouflaged with dirt.

RIGHT | A male *Viridasius* at Nosy Be, Madagascar. These nocturnal hunters have independently developed a special type of eyes, equipped with a reflective layer in the shape of a grille.

GENUS
Viridasius

DISTRIBUTION
Madagascar and the Comoro Islands

HABITAT
Tropical rainforests

CHARACTERISTICS

- Two claws and claw tufts
- Eyes in three rows; posterior eyes reflective, with grate-like tapeta
- Highly contrasting, disruptive coloration

DESERT WOLF SPIDERS

Xenoctenids are a small, recently recognized family restricted to the neotropics. They resemble ctenids and miturgids, but molecular data place them as a separate lineage. Xenoctenids are vagrant spiders that live on the soil. As in many other nocturnally active hunters, they possess a grate-shaped tapetum that gives a characteristic shine to the eyes when these are illuminated with a headlamp.

Xenoctenus are rather large spiders that live in deserts and arid areas in central Argentina. They are extremely fast, and their long legs and striped colors make them very hard to spot when they stand still on the ground. The spiders' long claws are probably adapted for walking on sand and loose soil.

LEFT | A *Xenoctenus* from Valle de la Luna, Argentina. These spiders are most frequent in sandy soils, where they are perfectly camouflaged. Males are especially long-legged—the large species can spread to the size of an open hand.

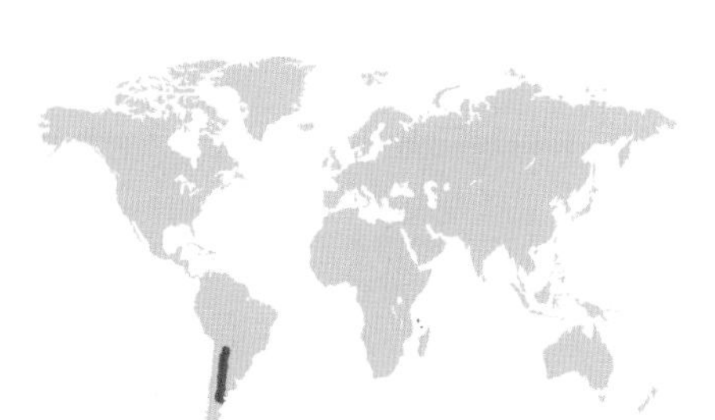

GENUS
Xenoctenus

DISTRIBUTION
Central Argentina

HABITAT
Deserts and arid areas; adapted to sandy soil

CHARACTERISTICS
- Two very long claws and projecting scopula
- Eyes in two recurved rows
- Secondary eyes reflective, with grate-like tapeta

CTENIDAE: *CTENUS*

TROPICAL WOLF SPIDERS

Ctenus is a genus of large spiders. They are characteristic of tropical rainforests, where they are very common although rarely noticed, due to their strictly nocturnal habits. The secondary eyes of these spiders reflect light (a function of the tapetum) and can thus be spotted at night using a headlamp. Even a small juvenile of about 0.2 inches (5 mm) in length can be seen at a distance of more than 10 yards (9 m). During the day, the spiders hide in the leaf-litter layer or in small holes in the ground, from which they emerge immediately after sunset.

Ctenids are categorized as hunters, but their strategy is primarily one of "sit and wait." They sit on their preferred substrate, , which varies from dead leaves, bare ground, or dead wood, depending on the species. Here, the spiders wait for passing prey, which may range from medium-sized insects to small frogs. They overpower the prey by sheer force, and subsequently bite and inject with venom.

Ctenus also offers an illustration on how systematics progresses. Hundreds of species from all over the world had been attributed to the genus

LEFT | Representatives of the Ctenidae are among the largest araneomorph spiders; females of the genus *Ctenus* carry their egg case in the chelicerae until they hatch.

RIGHT | Most tropical wolf spiders are well camouflaged but some display flashy colors when threatened; *Phoneutria nigriventer* even assumes an aggressive stance in such circumstances.

GENUS
Ctenus

DISTRIBUTION
Rainforests in Panama and South America

HABITAT
Tropical rainforests

CHARACTERISTICS

- Eye arrangement in three rows: 2–4–2
- Secondary eyes reflect light with the grate-shaped tapetum.
- Two claws, claw tufts, and scopulae
- Nocturnal hunters in rainforests

on the basis of superficial morphological similarities. However, recent studies, also drawing on molecular data, have shown that *Ctenus* is actually restricted to the neotropics and the species from other parts of the world belong to other genera.

The family Ctenidae also contains one of the few genuinely dangerous spiders, *Phoneutria nigriventer.* This large ctenid is common in urban areas of Brazil, and is aggressive, even toward humans—a behavior that is highly unusual. Because its venom is neurotoxic, targeting the nervous system, a bite from this animal may be life-threatening; an antivenom is available. Because one side effect of a bite can be an erection in males, the venom's many components are under study as a potential treatment for impotence.

OXYOPIDAE: *PEUCETIA*

GREEN LYNX SPIDERS

Peucetia is a widely distributed genus that includes more than 50 sizable species. With their primarily flashy green color and black, yellow, and reddish markings, they are among the most handsome spiders. They are plant dwellers, preferring to sit on plants with glandular hairs; they then profit off the insects that are immobilized by those hairs. To avoid contact with the sticky leaves, the spiders prefer to walk on the threads of their own silk, which they spin over them.

The name "lynx spiders" refers to another predatory strategy that these spiders have adopted: they jump on their prey, sometimes from the top of a plant, catching it in midair. The long, erect spines help to immobilize the prey, while at the same time keeping it at sufficient distance to avoid a bite or a sting.

Male *Peucetia* perform simple dances in front of the female during courtship, implying that their eyesight is better than that of the average spider. The couple usually hangs from a silk thread during copulation. Such a position is particularly rare among spiders, but it may be a way of avoiding contact with their preferred sticky substrate.

LEFT | It is not clear why some lynx spiders of the genus *Peucetia* have bright colors when their method of ambushing flying insects leaves them exposed during daylight on the tops of shrubs. It is possible that their disrupting colors misguide predators in search of spider prey.

RIGHT | The Green Lynx Spider (*Peucetia viridans*) is well camouflaged when it sits on its preferred substrate—short shrubs with sticky hairs—but it is particularly eye-catching on a bright flower.

GENUS
Peucetia

DISTRIBUTION
Cosmopolitan (in tropical and subtropical areas)

HABITAT
Low shrubs, preferably with glandular hairs, in open habitat

CHARACTERISTICS
- Hexagonal eye pattern
- Legs with many long, erect spines
- Green carapace and tapering abdomen

BARK HUNTERS

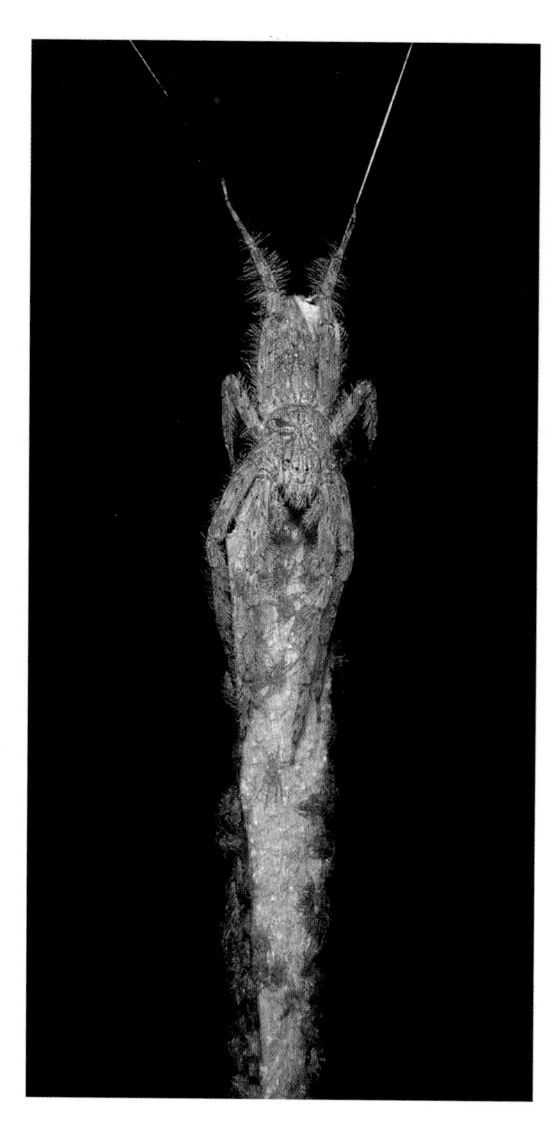

Senoculids are a small family comprising a single genus, *Senoculus*. There are about 30 known species, all of which live in rainforests in the Americas. Their spotted coloration and the long hairs on their legs make them extremely cryptic in their typical habitat, on tree bark. Gravid females hang a dry leaf from two silk threads, attach an egg sac to it, and hide the sac from view with their own bodies.

Senoculus rest on the underside of branches during the day, and hunt during the night. Their posterior eyes are equipped with grate-shaped tapeta. They are related to wolf spiders (Lycosidae) and lynx spiders (Oxyopidae), although taxonomists are still uncertain of their closest relative.

LEFT | A female *Senoculus* from Roraima, Brazil. The spider is guarding her egg sac, from which the spiderlings are just emerging. The eggs are placed on a dry leaf hanging from two silk lines, and the female camouflages them with her own body.

GENUS
Senoculus

DISTRIBUTION
America, from Mexico to northern Argentina

HABITAT
Rainforests

CHARACTERISTICS

- Eyes in three rows, the anterior laterals very small; posterior eyes with grate-shaped tapeta
- Three claws
- Very cryptic

AMERICAN FISHING SPIDERS

Trechaleids are mainly American spiders, although a few species from Asia may also belong in the family. They specialize in hunting over the surface of streams and are aided in this by water-repellent hairs. These allow them to walk on water, or take a dip to hunt swimming prey, and emerge from the water perfectly dry.

As part of their courtship, male trechaleids may offer an insect wrapped in silk as a nuptial gift to the female; mating takes place while she is busy consuming the gift. When prey is scarce, males may offer a stone wrapped in silk—a worthless gift that still enables them to mate. The females produce a lentil-shaped egg sac and carry it attached to the spinnerets, as wolf spiders (Lycosidae) do, but the newly emerged spiderlings remain on the egg sac, rather than on the female's abdomen. As their egg-sac-carrying behavior suggests, trechaleids are closely related to wolf spiders. Some species have an impressively large leg span, as wide as an open hand, even though their bodies are narrow.

The aquatic trechaleid species are nocturnal; they keep their front legs extended on the water surface, while holding onto a stone or log with their posterior legs. They detect and hunt insects, or even fish, by vibrations on the water surface.

Trechalea has nine species, which live on streams in tropical forests in America.

LEFT | A female *Trechalea* from Panguana Nature Reserve, Peru. Female trechaleids carry their egg sacs with their spinnerets, as wolf spiders do. This allows them to move around streams and hunt prey without leaving the eggs unattended.

GENUS
Trechalea

DISTRIBUTION
America, from Mexico to northern Argentina

HABITAT
Streams in tropical forests

CHARACTERISTICS
- Eyes in two rows; posterior eyes with grate-shaped tapeta
- Walks on water
- Females carry their egg sacs on their spinnerets

PISAURIDAE: *DOLOMEDES*

FISHING SPIDERS

Spiders of the genus *Dolomedes*, which includes slightly more than 100 described species, build small webs only as juveniles. Later, they hunt near open water without building a web, relying instead on the surface to conduct vibrations. The spiders hold the tips of their leg tarsi on the water surface. They can detect vibration by sensing even the tiniest deflection of the tarsi against the preceding leg segment, the metatarsus. Tiny slit sensilla, with nerves ending in a membranous cover on the cuticle, alert the central nervous system to changes in the environment. Wind, for example, is filtered out, by the low frequency of the signal. Insects, in contrast, produce a signal that triggers prey-capture behavior in the spider: it runs over the water surface in the direction of the insect, injects venom through its fangs to immobilize the prey, and starts consuming it by regurgitating digestive fluid through its mouth and chewing prey and fluid with its chelicerae. After the digestible parts of the prey become liquefied, the spider sucks the nutrients into its stomach. *Dolomedes* are adept predators that

LEFT | A Raft Spider (*Dolomedes fimbriatus*) feeding on a freshly caught stickleback. Spiders can detect vibrations not only from the water surface, but also from fish and tadpoles swimming in the water. Their ability to dive enlarges their prey spectrum.

RIGHT | A female Raft Spider (*Dolomedes fimbriatus*) holding her egg sac with her fangs. Note the typical eye arrangement and parchment-like outer layer of the egg sac. The palps are in close contact with this layer and can detect the slightest movements when the spiderlings hatch from their eggs.

GENUS
Dolomedes

DISTRIBUTION
Cosmopolitan (although only one species has been described from South America)

HABITAT
Near open water; uses the water surface for hunting

CHARACTERISTICS
- Hunts on the water surface and can dive below it
- Females carry the egg sac in their chelicerae
- Light lateral bands on carapace

can even catch small fish near the water surface (hence their common name of fishing spiders).

The egg sac of *Dolomedes* and other pisaurids is a large, spherical ball, held primarily by the fangs. Only the outer silken layers are pierced. When hatching begins, the mother will cut into the parchment-like outer layer of the egg sac to help the spiderlings emerge.

Other pisaurids have diverse lifestyles. Some build webs to catch prey, some run on top of the web (e.g., *Dendrolycosa*), whereas others hang underneath (e.g., *Sphedanus*, *Euprosthenops*). Some other genera (e.g., *Pisaura*) hunt their prey without using a web, but do build a nursery web to protect their newly hatched spiderlings. Males of some *Pisaura* species are notable for using nuptial gifts as part of their courtship behavior—catching and wrapping a prey item that is presented to the female for her to eat while he is busy mating with her. In some *Dolomedes* species, the males die soon after initiating copulation, and are typically cannibalized by the females, thereby providing extra nutrition for laying the eggs.

LACE SHEET-WEAVERS

RIGHT | A female *Psechrus* hanging in her web and holding her egg sac with her fangs. Note the characteristic banded leg pattern and the ornamented body coloration.

ABOVE | A male *Fecenia cylindrata* showing its typical leg position, used for hiding in a rolled leaf in the center of its pseudo-orb web.

Every tourist who has entered the jungles of Asia has likely encountered the large, dome-shaped webs of *Psechrus*. Although these are usually built in forests close to trees, logs, or rock crevices, or in caves, some individuals will enter buildings to spin their web. The only requirement the spider needs is an accessible retreat to hide in during daytime. These shy spiders come out at night, hanging under the huge web and waiting for prey. The web consists of sturdy threads that act as a functional frame, and of cribellate threads that are applied in parallel lines across the frame threads. The cribellate threads are good at capturing insects because small structures—such as insect setae or claws—are easily entangled in the micro-fine mesh of the silk.

Psechrus are reclusive, and not much is known about their biology. They need shady places with relatively high humidity. The male uses a special leg position during courtship: the four forelegs are

GENUS
Psechrus

DISTRIBUTION
Tropical and subtropical Asia

HABITAT
Shady, humid places—usually forests, among trees, and in rock crevices, and caves; some species adapt to man-made structures

CHARACTERISTICS
- Build large, dome-shaped webs
- Underside of abdomen has a light median line
- Females hold egg sac in fangs

widely spread. After a successful mating, the female will lay about 70–170 eggs in a spherical egg sac, which she carries in her fangs. All 57 described species have a similar color pattern, with a characteristic light median line on the underside of the abdomen.

The second genus in this family, *Fecenia*, contains four species that build a very different web. Their two-dimensional pseudo-orb web is constructed in the vegetation of rainforests, with the spider hiding during the day in a rolled leaf hanging in the center. At night, it extends only a few legs from the retreat until prey is detected. Interestingly, juveniles build a small, dome-shaped web that resembles the large webs of *Psechrus*.

LYCOSIDAE: *PARDOSA*

COMMON WOLF SPIDERS

Lycosidae, the family to which the genus *Pardosa* belongs, includes almost 2,500 species, making it one of the largest spider families. Lycosids are also among the most frequently spotted spiders since they are fairly large, are active during the day, and often form dense populations in open habitats. Grasslands appear to be the wolf spiders' preferred environment. The family has seen a large recent radiation, apparently in parallel with the worldwide spread of grass-dominated habitats since the Miocene (14 million years ago). This explains the existence of many closely related species in the large genus *Pardosa*, which contains almost 550 species that are often difficult to separate.

Pardosa, and by extension most wolf spiders, are ground-living hunters par excellence. Yet, they lack some of the most characteristic adaptations of wandering spiders: they do not have claw tufts and are only rarely provided with scopulae (specialized hairs under their tarsi that assist with grabbing and holding onto prey). Instead, the presence of a third tarsal claw betrays their recent descent from web-building spiders. It also implies that their activity is largely restricted to the ground

LEFT | The Spotted Wolf Spider (*Pardosa amentata*) is the most common representative of the Lycosidae in western European gardens; the male uses the bright morning sun in spring to display its courtship dance.

RIGHT | The spiderlings of this *Pardosa morosa* emerge from the egg case and climb onto their mother's back, clinging to special hooked hairs on the dorsum of her abdomen.

GENUS
Pardosa

DISTRIBUTION
Cosmopolitan

HABITAT
Grasslands and fields

CHARACTERISTICS
- Eyes arranged in three rows: 4–2–2
- Good vision
- Courtship includes species-specific performance by males
- Females carry egg sac on spinnerets and spiderlings on upperside of abdomen

and herb layers, since they lack leg specializations to climb steep surfaces.

Wolf spiders have sharp vision, a capacity reflected in the courtship behavior of males. They perform species-specific dances, often including waving of their contrastingly colored palps and legs. Some of the hundreds of species have been recognized by the details of these dances before they could be distinguished by subtle morphological differences.

Females carry the globular, pale gray egg sac attached to their spinnerets for several weeks. When the spiderlings are ready to hatch, the mother cuts a slit in the egg sac. The spiderlings then climb onto their mother's abdomen and remain there until after their next molt.

LYCOSIDAE: *OCYALE*

BEACH WOLF SPIDERS

Since the majority of wolf spiders are active by day and thus often exposed, they tend to be well camouflaged. Even the night-active representatives have cryptic colors because they live in habitats with few possible retreats. Those that inhabit grassland usually have longitudinal bands; other types of habitats house species with other kinds of color patterns.

Most species of *Ocyale* live on bare sandy shores of inland waters, and are clearly adapted to this type of substrate. *Ocyale ghost* is almost pure white: it is found on the white-sand beaches of an inland lake in Madagascar, and is thus perfectly camouflaged. Although protected to some degree by their color, these wolf spiders hide by day in a shallow, silk-lined retreat just under the sand surface. At night, they prey on insects and other arthropods (including conspecifics, a phenomenon that is not rare among wolf spiders). They are well equipped to subdue large, dangerous prey. Their fairly long legs, provided with erectile spines, help them overpower the prey while keeping most of their body out of harm's way. Only then is the prey bitten and injected with venom.

LEFT | White spiders are rather rare, but this *Ocyale ghost* from Madagascar is perfectly adapted to the white sand around the inland lake where it lives.

RIGHT | Camouflage did not help a pale grasshopper escape from this *Ocyale ghost*, which not only has good eyesight but can also detect the slightest movement.

GENUS
Ocyale

DISTRIBUTION
Old World tropics and subtropics

HABITAT
Sandy shores of inland waters with little vegetation

CHARACTERISTICS
- Color pattern plain, without bands
- Epigynum triangular, with a dense cover of thick hairs
- Lives on sandy beaches

LYCOSIDAE: *HIPPASA*

FUNNEL-WEB WOLF SPIDERS

Like other lycosids, members of the genus *Hippasa* have the characteristic wolf spider eye pattern: a row of four small eyes in front, followed by two consecutive pairs of larger ones. The grate-shaped tapeta of the secondary eyes reflect light, making them easily traceable at night in the beam of a headlamp. Members of the genus are unique in the family in having large posterior spinnerets consisting of two equally long articles. These are clearly linked with the construction of a web—in this case, a densely woven sheet with a funnel leading into the ground or dense herb layer. The spider runs on top of the sheet in a manner similar to that of the funnel-web weavers (Agelenidae).

Shortly after the breeding season, grasslands may be covered by tiny webs of the spiderlings that have left the protection their mothers. *Hippasa* is a perfect example of the original web-building habit of the Lycosidae, a lifestyle abandoned by the majority of species, which are free-roaming hunters.

LEFT | Wolf spiders of the genus *Hippasa* are often mistaken for funnel-web spiders of the family Agelenidae because of their similar lifestyle.

RIGHT | *Hippasa* species live in grassland on a sheet web that has a funnel-shaped retreat in the ground; web-building is probably primitive for the family.

GENUS
Hippasa

DISTRIBUTION
Old World tropics and subtropics

HABITAT
Grasslands, open places with short vegetation

CHARACTERISTICS

- Abdomen with two rows of white spots
- Posterior spinnerets much longer than anterior ones, with a long distal segment
- Lives in a funnel web at ground level

LAMPONIDAE: *LAMPONA*

WHITE-TAILED SPIDERS

RIGHT | A female *Lampona cylindrata* from Western Australia; the species is frequently found in and around buildings, and this female was inadvertently transported to Germany in a suitcase.

Lampona is the best known of the 23 genera currently placed in the Lamponidae. Its members are commonly called white-tailed spiders because of the conspicuous white spot at the rear of the abdomen, just above the spinnerets. One species, *Lampona cylindrata*, has gained considerable notoriety in Australia. It is one of the most frequently collected spiders there, as it is easily the largest species of the family and, in addition to occurring in native forests, is also synanthropic. Frequently found around houses, these animals fearlessly invade the webs of other spiders, preying even on the large *Badumna* species that occur in the same habitats.

Because of their synanthropic habits, white-tailed spiders are sometimes found in odd places (e.g., inside a plastic-wrapped head of lettuce). They are large and strong enough to break through human skin when they bite—something most spiders are too small and weak to do. Some toxicologists have

GENUS
Lampona

DISTRIBUTION
Australia, New Guinea, New Caledonia; introduced to New Zealand

HABITAT
Native forests and buildings

CHARACTERISTICS
- Pair of invaginated oval sclerites just behind the book lungs
- Tubular pedicel (in lamponines only)
- Distal article of anterior lateral spinnerets represented by an incomplete ring of sclerotized cuticle

suggested that *Lampona* bites are responsible for producing severe necrotic lesions, much like those of brown recluse spiders (*Loxosceles*). Investigating that accusation has been complicated by issues of specific identity. What had been considered *Lampona cylindrata* has turned out to be a complex of three closely related species (one in New Guinea, one found across southern Australia, and a third occurring along the east coast of Australia, more commonly in the north).

Curiously, both Australian species have been introduced to New Zealand; the northern Australian species occurs only on the North Island and the southern Australian species only on the South Island. Investigations of 130 cases where the biting spiders of the two Australian species have been definitively identified have revealed no examples of severe necrosis. Laboratory tests of the venom strength have been equally negative; actual bite reactions are not severe, and these handsome spiders appear to have been maligned unfairly.

ABOVE | A female *Lampona murina*, feeding on a desid spider (*Cambridgea*), on the North Island of New Zealand.

PRODIDOMIDAE: *PRODIDOMUS*

LONG-SPINNERETED SPIDERS

Prodidomidae is a substantial group, including three subfamilies, more than 30 genera, and more than 300 species. Its members have elongated spigots on their anterior spinnerets; each spigot is associated with long hairs. *Prodidomus* species have a distinctive eye pattern, with the posterior median and lateral eyes of each side forming a line behind the anterior lateral eyes. They are also characterized by their enlarged, divergent chelicerae with long, curved fangs.

Prodidomus was first described from the USA, but the genus is probably native only to the Old World. The species found in the USA is synanthropic, and has therefore turned up in diverse places, including Cuba, Argentina, Chile, Japan, China, New Caledonia, and St. Helena.

In the subfamily Molycriinae, the anterior spinnerets are even more unusual: they originate near the front, rather than the back, of the abdomen, and occupy almost its entire length.

LEFT | An anterior view of a Mediterranean species of *Prodidomus*. The eye arrangement, with the large, circular anterior median eyes flanked by the obliquely oriented rows of other eyes, is characteristic of this genus.

GENUS
Prodidomus

DISTRIBUTION
Mediterranean to Japan, Africa, Australia; introduced in the New World (USA to Argentina and Chile)

HABITAT
Under rocks in dry areas

CHARACTERISTICS
- Anterior median eyes round, other eyes oval, forming a triangular pattern
- Chelicerae divergent, with long fangs
- Anterior lateral spinnerets with elongated spigots

How the animals use these bizarre spinnerets is unknown. However, it appears that they are independently mobile, which may allow the spiders to spin a wide swathe of silk in a semicircular pattern.

Molycriines are primarily Australian, but occur as far north as islands near Malaysia. Oddly, the extraordinarily long anterior spinnerets are shared with one genus, *Zimiris*, of the subfamily Prodidominae, which contains only two species. These are also each synanthropic and hence widely introduced to countries around the globe.

ABOVE | An Australian *Molycria*; the long, slender legs of these species enable them to run quickly.

ABOVE, INSET | A ventral view of an Australian molycriine; note the greatly elongated anterior lateral spinnerets, which are characteristic of this subfamily.

GNAPHOSIDAE: *EILICA*

GROUND SPIDERS

Gnaphosids are one of the largest families of spiders, with more than 2,200 species worldwide. Some of the most interesting belong to the subfamily Laroniinae, members of which differ from other ground spiders in having flat, lobe-like laminae, rather than teeth, on the rear margin of the chelicerae. *Eilica* has an extensive distribution, being found in tropical and subtropical America, Africa, Australia, and India—all areas once part of Gondwana, the southern supercontinent that began to separate into today's continents some 180 million years ago.

Although laroniine gnaphosids are not ant mimics, they seem to have close connections with ants. For most spiders, ants are too dangerous to prey upon, but some *Eilica* species live under rocks,

RIGHT | In common with *Eilica* species, this elegant female *Callilepis* from Austria is associated with ants and can even feed on them.

LEFT | This female *Eilica* from South Africa shows the nicely patterned abdomen; other members of the genus found on other continents often have similarly attractive abdominal color patterns.

GENUS
Eilica

DISTRIBUTION
Southern USA to Argentina, Africa, India, Australia

HABITAT
Near ant nests in wet and dry forests

CHARACTERISTICS
- Chelicerae with flat lobes opposite rear side of fangs
- Anterior lateral spinnerets tubular, widely separated
- Some spigots on anterior lateral spinnerets greatly widened

together with carpenter ants. There they construct tough, pill-shaped egg sacs; after hatching and biting through the egg sac to escape, the spiderlings freely crawl down the ant nest's entry holes, and are ignored by the worker ants they encounter. The spiderlings return to the empty egg sacs and use them as retreats. When an ant nest is disturbed, the worker ants move all the egg sacs into the interior of their nest, where they are protected from predators and parasites. The spiders clearly benefit from access to the favorable environments occupied by the ants in the dry environments in which they occur. They have thus presumably evolved chemical similarities that allow them to "pass" as ants.

The likely closest relative of *Eilica* is the genus *Callilepis*. This does not occur in Gondwanan areas, but rather in North America, Europe, and Asia north of India—all areas that belonged to Laurasia, the supercontinent from which Gondwana separated (prior to that, they were united in a huge supercontinent, Pangaea). The origin of the two genera thus probably dates to the division of Pangaea.

At least some species of *Callilepis* have moved beyond coexisting with ants, having developed a technique for feeding on these dangerous prey. The prey ant is always attacked from the front, and the spider's front legs are used to locate the base of one of the ant's antennae. The spider bites there quickly (0.2 seconds) and then beats a hasty retreat. Although the prey ant is initially fierce, within a few seconds the bitten antenna becomes limp and the ant is able to move only in small circles, enabling the spider to relocate it easily. The spider moves the injured ant back to its retreat, and closes the retreat off with silk while it feasts.

AMMOXENIDAE: *AMMOXENUS*

TERMITE HUNTERS

Ammoxenus are specialist hunters on termites, and are found on sand dunes in arid regions, as well as in savannas and tropical bushlands in southern Africa—usually in association with harvester termites. When the termites are actively foraging, the spiders are also active and agile on the surface, darting quickly between grass clumps; when the termites are not active, the spiders bury themselves in sandy soil near the termite foraging holes, or even inside the soil mounds made by the termites. Inside the mounds, they construct close-fitting, sac-shaped retreats composed of an inner layer of shiny silk and an outer layer of sticky silk to which sand grains adhere.

The spiders are sometimes called sand-divers, as they can use the modified setae on their robust chelicerae to dive headfirst into the sand. As they disappear under the sand, they adopt an upside-down posture about 0.5 inches (13 mm) below the surface. Even the juveniles can jump onto worker termites as large as they are, biting them just behind the head capsule. The termite dies quickly, within 30–65 seconds, apparently before it can emit alarm pheromones. The spider then dives under the sand, burying itself and at least partially burying the termite as well, while feeding on it from below. The spiders sometimes wrap parcels of four to eight dead termites in silk, apparently for use as a food reserve for when the termites are not foraging.

LEFT | This *Ammoxenus* is diving headfirst into the sand; within a few seconds, the entire animal will be under the surface.

GENUS
Ammoxenus

DISTRIBUTION
Namibia, Botswana, South Africa

HABITAT
Sand dunes in arid regions; also savannas and tropical bushlands (close to harvester termites)

CHARACTERISTICS
- Chelicerae project forward, with spines used for digging
- Leg tarsi pseudosegmented, flexible
- Feed on termites; adept at digging in sand

CITHAERONIDAE: *CITHAERON*

CURLY-LEGGED GROUND SPIDERS

BELOW | Even though the legs of this *Cithaeron* are so thin, these spiders can run with amazing speed.

Cithaeron include some of the fastest ground-dwelling spiders. They are most often found in silk retreats under stones or piles of grass, but run so rapidly when disturbed that even experienced collectors have difficulty catching them. Males have been found sharing retreats with subadult females, presumably so that they can mate as soon as the females undergo their final molt and become adults. As in *Ammoxenus*, the long leg tarsi are pseudosegmented, making them extremely flexible; in preserved specimens, the tarsi curl (hence the common name, curly-legged ground spiders).

The most common species is widespread, apparently occurring naturally from Greece and North Africa through the Middle East and India to Malaysia. Interestingly, it has apparently been introduced to Florida in the USA, Cuba, Brazil, and even Australia. At least in Florida, the spiders are synanthropic, found both inside houses and on their outside walls. They have been observed feeding on other spiders, and that may be their preferred food source.

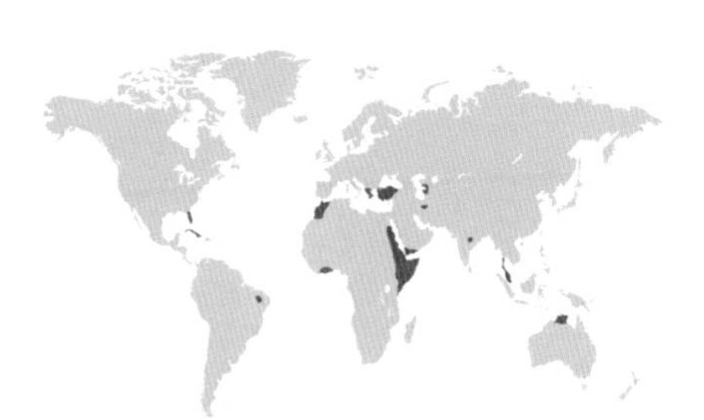

GENUS
Cithaeron

DISTRIBUTION
Africa, Eurasia from Greece to Malaysia; introduced to Florida, Cuba, Brazil, Australia

HABITAT
Ground-dwelling among grassy clumps and stones; most common species synanthropic

CHARACTERISTICS

- Long, narrow legs with pseudosegmented, flexible tarsi
- Posterior median eyes flattened and irregularly shaped
- Posterior median spinnerets with numerous small spigots
- Chelicerae lack teeth

GALLIENIELLIDAE: *GALLIENIELLA*

LONG-JAWED GROUND SPIDERS

Gallieniella species are known only from Madagascar and the nearby Comoros, but closely related genera occur in South Africa, Argentina, and Australia. Males of *Gallieniella* are striking in appearance; the first known species was described as *Gallieniella mygaloides*, because of its mygalomorph-like appearance. The chelicerae are almost as long as the carapace, extending far forward, and bear long fangs that extend backward. In females, the chelicerae and fangs are not quite as impressively modified, but are still longer than in most other ground spiders.

Gallieniella spiders have been found running with ants of similar appearance and move with such agility that they are difficult to distinguish from the ants—and to capture. The long chelicerae and fangs probably allow these spiders to capture ants while keeping the rest of their bodies safely away from the prey.

RIGHT | A female of the Argentine species *Galianoella leucostigma*, the only gallieniellid described from the New World; note the anteriorly directed chelicerae.

GENUS
Gallieniella

DISTRIBUTION
Madagascar, Comoro Islands

HABITAT
Litter, moss, and soil in montane rainforests

CHARACTERISTICS

- Chelicerae greatly elongated, extending forward
- Fangs long, extending almost longitudinally
- Posterior median eyes small and flattened

SCORPION SPIDERS

Platyoides and their relatives are spiders that look like someone has dropped an unabridged dictionary on them. Their bodies are greatly flattened, enabling them to fit into narrow crevices under bark or between rocks. Even their legs are laterigrade—twisted so that the morphologically front surface faces up, thus reducing the effective height of the legs. The trochanters on the posterior pair of legs are greatly elongated, allowing the femora of those legs to extend posteriorly, along the sides of the abdomen or even folding over it.

One of the South African species can be found in houses and has even been introduced to Australia. Other trochanteriid genera are found in the southern hemisphere, and from India to Japan.

LEFT | A female scorpion spider from South Africa; the extremely flat body and rotated legs allow trochanteriids to inhabit very narrow crevices.

GENUS
Platyoides

DISTRIBUTION
Canary Islands, southern and eastern Africa, Madagascar, Réunion, Seychelles

HABITAT
Forests and rocky areas, in crevices of tree bark and rocks; one South African species is synanthropic

CHARACTERISTICS
- Greatly flattened body
- Divergent chelicerae
- Posterior trochanters elongated

INTERTIDAL GHOST SPIDERS

BELOW | A female *Amaurobioides maritima*, from the South Island of New Zealand. These spiders live exclusively on rocks at the seashore, and can withstand immersion during high tide. The dense silk of their retreat is watertight.

The 12 species in the genus *Amaurobioides* exploit a highly unusual habitat for a spider: they spin dense silk retreats in crevices on the rocky seashores of South Africa, southern mainland Australia, Tasmania, New Zealand, and Chile. They seal the retreat to endure immersion during high tides. During the night, the spiders hunt from the cell entrance, ambushing passing isopod and amphipod crustaceans.

Genetic studies suggest that *Amaurobioides* originated from a Patagonian lineage that dispersed to South Africa, where they developed their adaptations to an intertidal lifestyle about 8 million years ago. In the South African species, *Amaurobioides africana*, the retreats are constructed near the base of large kelp seaweeds. These algae form huge rafts that circle the southern oceans on the circumpolar current and the west wind drift.

GENUS
Amaurobioides

DISTRIBUTION
Rocky coasts of southern continents, Auckland Islands, Campbell Islands

HABITAT
Intertidal, on rocky seashores (webs spun in rock crevices, and become submerged at high tide); one South African species lives close to large kelp seaweeds

CHARACTERISTICS
- Extended tracheal system, passing to the cephalothorax and into the legs
- Feet with spatulate adhesive setae
- Makes a silken retreat in rock crevices that can survive immersion during high tide

Helped by the sea currents and their ability to withstand immersion, *Amaurobioides* probably colonized and diversified in the coastal habitats of the southern continents, leaving a trail in the species phylogeny: the oldest group in South Africa, then in Australia, later in New Zealand, and, finally, a long-range dispersal from New Zealand to Chile. This closes a journey around the world that took around 8 million years, as the species' distant, ancient origins were in the New World.

The other members of the family Anyphaenidae are mainly South American. Most species are nocturnal hunters that wander in the foliage and catch their prey without the aid of webs. Their common name, ghost spiders, reflects their long, pale legs, which blur when the spiders flee at high speed.

TOP | A female *Amaurobioides litoralis*, from Tasmania. All *Amaurobioides* species are very similar, having diverged relatively recently from a South American ancestor and dispersed around the austral sea coasts.

ABOVE | A silken cell of *Amaurobioides pleta*, New Zealand.

LIOCRANIDAE: *HORTIPES*

GARDEN LEGS SPIDERS

Hortipes is a large genus—including more than 70 species—of tiny spiders that are often called "funny eyes": they have a strange, cross-eyed look because the dark retina is eccentric in relation to the lens above it. They have earned their other name, "garden legs," from the presence of an oval patch of curved hairs arranged around a trichobothrium on the metatarsi of the first two

RIGHT | The tiny *Hortipes* are well equipped to overpower small, fast-moving prey: they not only have a special device to detect prey, but the first leg pairs are provided with numerous spines to enclose their victims.

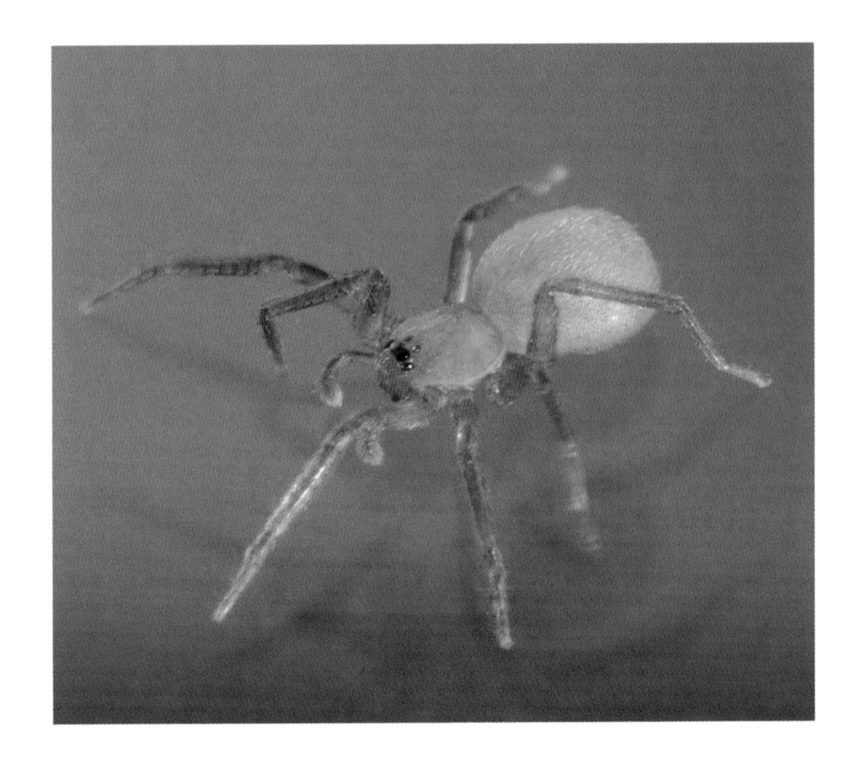

GENUS
Hortipes

DISTRIBUTION
Tropical Africa

HABITAT
Leaf litter of humid habitats, in lowland to montane forests

CHARACTERISTICS

- Metatarsi of front legs have "tiny gardens": a trichobothrium surrounded by an oval "fence" of curved hairs
- Double rows of ventral spines on the femora, tibiae, and metatarsi of legs I and II
- Anterior median eyes have an odd appearance, because the dark retina occupies only one side of each lens

leg pairs. It is assumed that these are directional air-motion sensors, the lateral setae shielding signals from the sides and behind. The hairs seem to provide these spiders with the necessary information on the whereabouts of collembolans (springtails), their preferred prey.

Hortipes has a broad distribution across Africa. Members of the genus are present in wetter habitats where some leaf litter has accumulated, from lowland forests to montane thickets. Strangely, though, they are not found in the huge miombo woodland in the southern half of the continent, even though those habitats have a well-developed leaf-litter layer that houses many collembolans.

The taxonomic position of the genus has shifted several times. It was initially described in Liocranidae but is sometimes placed in the Corinnidae, although its morphology is certainly not characteristic of that family.

Representatives of *Agroeca*, another genus in the Liocranidae, are larger. The so-called "fairy lamp spiders" have a Holarctic distribution and occupy a wide variety of habitats, from heathland with low shrubs to open woodland and forests. *Agroeca* species are nocturnal hunters, roaming slowly over the soil in search of ground-dwelling insects and other invertebrates, including woodlice. Due to their cryptic lifestyle these spiders are rarely observed, but their presence is given away by the remarkable egg case some of them construct. This is a conspicuous, bright white barrel-shaped container, hanging from a stalk under a twig or other element of the vegetation. It resembles a lantern in shape, hence the vernacular name of these spiders. Once the construction is complete, it is entirely covered with mud, rendering it much less eye-catching.

BELOW | The shape of the bright white egg case of *Agroeca brunnea* resembles a lantern but is eventually covered with dirt to make it less conspicuous.

BELOW RIGHT | The anterior legs of *Agroeca brunnea* have several pairs of ventral spines, typical equipment for hunters that rely on their strength to catch prey.

CLUBIONIDAE: *CLUBIONA*

SAC SPIDERS

Clubiona is a large, cosmopolitan genus that includes almost 500 species of hunting spider. They are typically pale yellowish or brownish with long spinnerets and contrasting dark chelicerae; the chelicerae are often longer in males. Most species live in lower vegetation, shrubs, or grasses—their preferred hunting grounds. They roam around during the night and spend the day in a densely woven, tubular silken retreat with an opening on both ends. These sacs are usually constructed under bark, in rolled leaves, or in other places well protected against rain.

These hiding places are also used for molting, mating, and depositing egg sacs, which are guarded until the eggs hatch. This behavior is apparently important, because feeding on unattended eggs of conspecifics has been observed in these spiders. The vacant nests of subadult and adult females appear to be impregnated with contact sex pheromones, eliciting courtship behavior from passing males.

Some *Clubiona* species live in low trees and are therefore considered useful predators for biological control, for instance in avocado plantations. They rest in the vicinity and have been observed to prey on caterpillars, often the main pest organisms—a behavior not common in spiders.

LEFT | A pale brown body with contrasting black chelicerae is the typical pattern of sac spiders of the genus *Clubiona*, of which *Clubiona lutescens* is an excellent example.

RIGHT | Speed is not needed for night hunters; they slowly close in on their prey and pounce in a short strike. This efficient hunting strategy is employed by *Clubiona* species.

GENUS
Clubiona

DISTRIBUTION
Cosmopolitan

HABITAT
Low vegetation, shrubs, and grasses; some species live in trees; retreats are constructed under bark or in rolled leaves

CHARACTERISTICS
- Yellowish to brown carapace
- Dark, swollen chelicerae
- Strong claw tufts

CORINNIDAE: *COPA*

SWIFT SAC SPIDERS

Copa is a small genus (with only seven species) belonging to the family Corinnidae (which currently contains fewer than 800 species). *Copa flavoplumosa* is very common in tropical and southern Africa. It is represented in almost all surveys of habitats with some leaf litter, ranging from dense woodland, over open savanna and fynbos, to semiarid areas with dispersed shrubs, but it is very rare in rainforest. Most species are primarily active by day and very fast hunters, with strong claw tufts.

Copa flavoplumosa is a remarkable example of a species with different color morphs. The most widespread variation has a bright orange carapace and an abdomen with black markings, the second a yellow carapace with black markings and a black abdomen with white markings, and the third almost completely black with white markings. It has been hypothesized that these are adaptations to local environmental conditions. However, the presence of the three forms in single populations in several areas in West Africa points to the possibility that individuals may be able to modify color during their development in response to microhabitat changes.

LEFT | Apart from the Lycosidae, *Copa* species are among the few spiders that hunt during the day. They are fast runners to avoid the risk of being preyed upon themselves, rather than to outrun their own prey.

RIGHT | *Copa flavopilosa* exhibits a remarkable color variation whose extent is rarely encountered in the world of spiders.

GENUS
Copa

DISTRIBUTION
Tropical and subtropical Africa, Madagascar, Sri Lanka, eastern Australia

HABITAT
Requires some leaf litter but can survive in a range of habitats: dense woodland, open savanna and fynbos, semiarid areas with patchy vegetation. Rare in rainforest

CHARACTERISTICS
- Eyes in a circular arrangement
- Markings covered with fine, plumose setae
- Fine proximal and distal dorsal setae on the anterior patellae

PHRUROLITHIDAE: *PHRUROTIMPUS*

GUARDSTONE SPIDERS

Phrurotimpus is a North American genus, but a closely related genus, *Phrurolithus*, occurs throughout Eurasia. That name, *Phrurolithus*, is a compound adjective meaning "guarding the stone." Its original describer, C. L. Koch, gave the spider that name in 1839 because "Her abode ... is always under stones. Underneath them the female lays her eggs in a spherical lump and spins over them a light, transparent web, on which the mother awaits the hatching of her young [translated from the German original description]." In the North American *Phrurotimpus*, those egg sacs are conspicuous—they are covered with silk that is dark pink to bright red. These spiders also have strong color patterns on both the carapace and abdomen.

There are two groups of *Phrurotimpus* species. In one group, the top and sides of the abdomen are dark with light markings; in the other, the top and sides of the abdomen are white, with dark markings. These genera, and the other phrurolithids, have several pairs of long spines underneath the tibiae and metatarsi of the front two pairs of legs.

LEFT | A female *Phrurolithus* species from Europe. Females of this genus attach their egg sac to the underside of a stone and are typically found there guarding it.

GENUS
Phrurotimpus

DISTRIBUTION
Canada to central Mexico

HABITAT
Forest leaf litter; egg sacs typically laid on underside of rocks

CHARACTERISTICS

- Tibiae and metatarsi of first two pairs of legs have several pairs of long spines
- Few other leg spines
- Femur of male palp has distal modifications

TRACHELIDAE: *TRACHELOPACHYS*

CUSPULED SPIDERS

Like most trachelids, the large, strikingly colored species of *Trachelopachys* have no leg spines, but the distal segments of the front two pairs of legs bear rows of short, thick, black cuspules. Most trachelids have a reddish carapace, but the South American species of *Trachelopachys* have a distinctively dark, often jet-black carapace. In life, the black carapace often contrasts with strikingly reddish legs, which typically have dark bands.

Trachelopachys occur in a wide variety of habitats, but some species have specialized niches. One, for example, seems to occur only on coastal sand dune areas, called restingas, in southeastern Brazil. Most of the animals inhabiting dunes are nocturnal, but the *Trachelopachys* are active only during the day. The dunes do have some shrubby vegetation that can withstand the salinity, but the spiders occur primarily on open sand. They are active in both the dry and rainy seasons, but their activity ceases for about three hours during the hottest part of the day, when the temperature at the surface of the sand can reach as high as 140°F (60°C). The striking coloration of the animals makes them very conspicuous against the white sand, and probably warns potential predators that they are unpalatable or otherwise undesirable as prey.

RIGHT | This elegant female of *Trachelopachys* hails from Argentina. The blackened carapace sets members of this genus apart from most other trachelids.

GENUS
Trachelopachys

DISTRIBUTION
South America

HABITAT
Varies; some species are specialists, e.g., coastal sand dunes

CHARACTERISTICS
- Spineless legs
- Tibiae, metatarsi, and tarsi of front two pairs of legs usually have short, thick, ventral cuspules (modified setae)
- Carapace black, legs red or light with dark rings

SELENOPIDAE: *SELENOPS*

FLATTIES

If there were a spider competition for making oneself invisible, *Selenops* would win in two categories: camouflage and speed. When these nocturnal spiders come out at twilight, they blend very well with their background—often tree bark. Both the body and the laterigrade legs stretching out in all directions show patterns and annulations that fit perfectly to the environment (laterigrade legs are slightly rotated, allowing the spiders to lie flat against the surrounding surfaces).

Despite their camouflage, these spiders can easily be found by directing a headlamp toward trees at night. The eyes of *Selenops* have an internal layer, the tapetum, that reflects the light, much like a cat's eyes; one can see their eyes shining like little diamonds at distances of 20 yards (18 m). Because of their very flat bodies, *Selenops* can hide in minute crevices. Their ability to rotate quickly is unsurpassed; they can spin their bodies in a complete circle more than eight times a second,

RIGHT | A female *Selenops muehlmannorum* from southern Laos, waiting on a rock to ambush prey at night. Note the characteristic laterigrade leg position.

GENUS
Selenops

DISTRIBUTION
Tropics and subtropics, except Australia

HABITAT
Tree bark and rock surfaces; sometimes in and around houses

CHARACTERISTICS
- Very flat body
- Eyes arranged in two rows: 6–2
- Hides in natural or man-made crevices

ABOVE | *Selenops* from Mozambique, showing the characteristic eye arrangement with all eight eyes close to the front margin of the carapace. The posterior lateral eyes are largest, and help detect potential prey movements at the sides and back of the spider.

and can strike at prey within 120 milliseconds. Because they can ambush prey and retreat with it in the blink of an eye, they are often welcomed as pest controllers in tropical houses.

With more than 130 described species, *Selenops* is the largest genus in the family; members of the other nine genera together include about as many species again. Most of the other genera are geographically more restricted; for example, the genera *Garcorops* and *Hovops* are known only from Madagascar and the adjacent Comoros.

SPARASSIDAE: *HETEROPODA*

FOREST HUNTSMAN SPIDERS

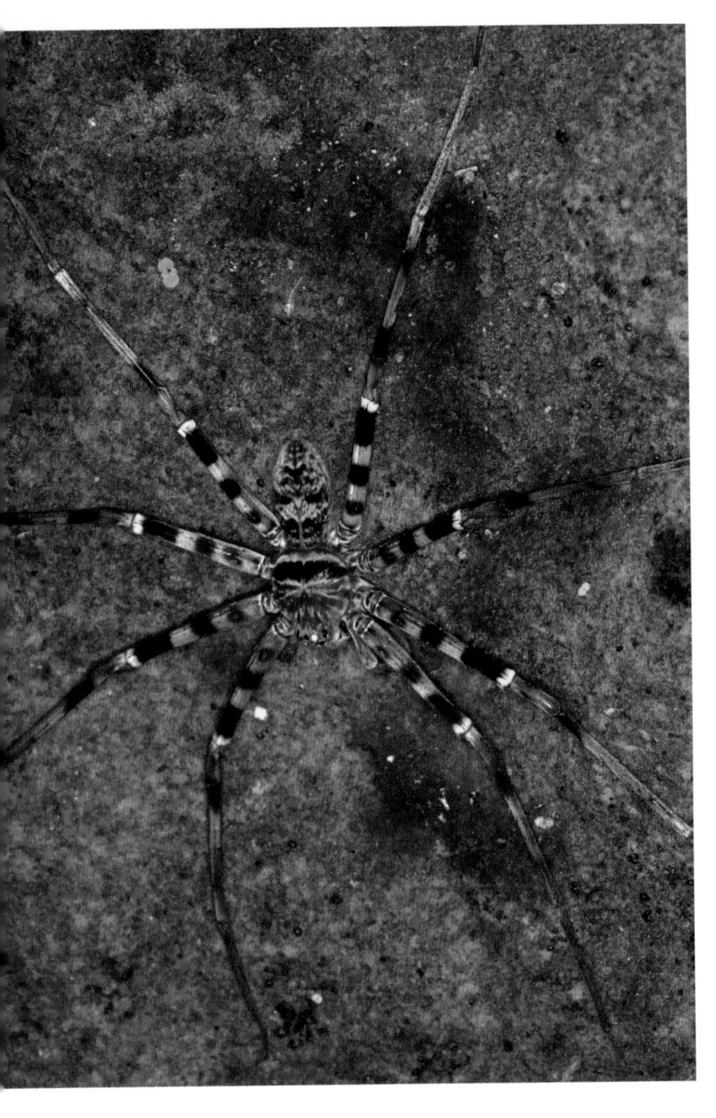

LEFT | A female *Heteropoda maxima* from Laos. This species has the largest leg span of the araneomorphs—up to nearly 12 inches (30 cm) in males.

RIGHT | A female *Heteropoda davidbowie* from Singapore, showing the typical folded legs and striking color pattern of the species.

Heteropoda species are distributed in tropical and subtropical rainforests, with several species occurring in caves. Around 200 species are described, ranging from 0.2 to 2 inches (5–50 mm) in body length; many more await scientific discovery. *Heteropoda* do not build webs for catching prey, but instead forage for prey at night. For these spiders, the eight eyes in two rows are not the dominant sense organs. Instead, the spiders rely on trichobothria—fine sensory hairs on the legs that detect even the slightest airborne vibrations. Together with slit sensilla (small innervated slits in the exoskeleton), this sensory instrumentation allows *Heteropoda* spiders to navigate easily in the dark.

GENUS
Heteropoda

DISTRIBUTION
Eastern Mediterranean, Afghanistan to Japan and Australia; one pantropical, introduced species

HABITAT
Tropical and subtropical rainforests, in leaf litter, on tree bark, or, rarely, on foliage; several species in caves; a few species in human settlements

CHARACTERISTICS
- Legs laterigrade (twisted sideways)
- Trilobed membrane at tips of leg metatarsi
- Fast-running spiders

Heteropoda davidbowie, described only in 2008, was named after the rock singer because of its striking coloration. In contrast to many other congeners, it lives in the understory foliage in rainforests from southern Thailand to Borneo. When sitting on smaller leaves, the spider has to fold its forelegs in order to fit the size of its chosen leaf. Due to its attractive appearance, it is one of the few spiders other than tarantulas that are often kept as pets. Males of the species were used in the Arachnid Orchestra Jam Sessions in Singapore's Centre for Contemporary Art; the drumming sounds they produce during courtship were greatly amplified to become the rhythm section. Other species occur in caves, including one of the largest known spiders, *Heteropoda maxima*—a species endemic to central Laos. It was discovered in 2001, first as preserved specimens in the scientific collections of the Natural History Museum in Paris, and two years later during an expedition to Laos that was featured in several television documentaries. A pantropical representative, *Heteropoda venatoria,* apparently originated in Southeast Asia, but has been introduced to many countries around the world. Even in the tropics, individuals of this species favor human settlements and buildings. They are also frequently found in cargo (e.g., bananas) being shipped around the world.

SPARASSIDAE: *CEBRENNUS*

DESERT HUNTSMAN SPIDERS

Most huntsman spiders live in humid environments; only a few species have adapted to arid habitats, especially deserts in Africa and Asia. *Cebrennus* spiders occur in deserts and steppes, excavating burrows in the ground or hiding under stones during the day. A few species build retreats in short vegetation. Rechenberg's Desert Huntsman (*Cebrennus rechenbergi*) was discovered in 2014 in a Moroccan desert; it is unique in escaping through a flic-flac-like movement. By this means, the spiders can move twice as fast as they would be able to run, and so have a better chance of escaping predators or parasitic wasps. The movement was analyzed by scientists and engineers, and used to help develop a robotic "spider" for potential Mars missions.

A closely related species, *Cebrennus villosus* from Tunisia and Algeria, does not exhibit this special

LEFT | A female *Cebrennus rechenbergi* raising its front legs as a defensive response. In the morning, tracks in the sand show where these spiders were active during the night.

RIGHT | *Cebrennus rechenbergi* from Morocco escaping by a flic-flac-like movement. With this special behavior the spider can survive attacks by parasitic wasps or larger predators.

GENUS
Cebrennus

DISTRIBUTION
Morocco and Canary Islands to Ethiopia and Afghanistan

HABITAT
Arid places in deserts and steppes; most species live in burrows, a few in retreats under stones or on low vegetation

CHARACTERISTICS
- Light coloration
- Kink in embolus of male palp
- Adapted to arid environments

escape behavior, but in common with its sister species it builds tubes in loose sand. Special setae on the palps and chelicerae enable the spiders of both species to carry sand from the bottom of the tube. Silk strands then fix the sand grains into a stable wall.

LEFT | A female *Cebrennus villosus* from Tunisia, digging its burrow and moving sand from the ground out of the tube. Rows of long, stiff setae on the pedipalps create a basket in which the spider can carry the sand.

PHILODROMIDAE: *PHILODROMUS*

RUNNING CRAB SPIDERS

Philodromus spiders look like small versions of huntsman spiders, but they lack the trilobed membranes at the tip of the leg metatarsi typical of the latter. More than 200 species of *Philodromus* have been described, from almost all regions of the world. Like sparassids, selenopids, and thomisids, they have laterigrade legs that are twisted to the sides. Philodromids were previously included in the family Thomisidae, but both morphological and molecular data show that they are not even the closest relatives of thomisids (and may be more closely related to jumping spiders). Unlike thomisids, their legs are all roughly the same length.

Philodromus do not build webs, but hunt their prey in foliage, on tree bark, or on the ground. Some species also occur on house walls. Others are well camouflaged in their environment, blending with the backgrounds on which they rest, such as sand dunes or lichens on tree bark, where their color pattern makes them almost invisible. *Philodromus* are among the spiders that can quickly autotomize a leg to escape a predator's attack; the leg is left in the predator's mouth but the spider escapes, with minimal loss of hemolymph. If the spider is not yet adult, a new leg can be generated in the next molt, and after a few molts it can even match the other legs in size.

Widespread philodromid species are flexible in their life cycle. In warmer climates only one year is needed to complete the life cycle, whereas in colder regions the same species need two years to mature and reproduce.

LEFT | *Philodromus aureolus* males have a shimmering appearance to their body; this was the basis for their specific name, which means "golden" or, figuratively, "beautiful."

GENUS
Philodromus

DISTRIBUTION
Mainly Holarctic, with several species known from the tropics

HABITAT
From forests to gardens, in foliage, and on tree bark, rocks, walls, and bare sand surfaces

CHARACTERISTICS
- Laterigrade legs of roughly equal length
- Eyes often ringed with white pigment
- Fast-running spiders

ABOVE | The Lichen Running Crab Spider (*Philodromus margaritatus*) occurs from Europe to Japan and blends perfectly with the surrounding lichens on tree bark.

THOMISIDAE: *MISUMENA*

FLOWER CRAB SPIDERS

BELOW | The yellow form of *Misumena vatia*, a species occurring in North America and from Europe to East Asia. These spiders can overwhelm insects that are much larger than them with a paralyzing bite to the head.

If one spider can be called the chameleon of all spiders, it is *Misumena vatia*. Females can change their color from white to yellow and back again, depending on the color of the flower on which they wait to ambush prey. The color change takes a few days and is caused by either production or destruction of yellow pigments called ommochromes inside the body. The white color is due to guanine crystals, produced by the spider's metabolism and often used for color patterns in other spiders as well (e.g., in the garden cross spider, a common orb weaver). By this means, the spiders can blend with the particular background they choose; they are therefore inconspicuous to potential predators and in some cases also to their prey. Another genus with the ability to change color is *Thomisus*. In some of these species the eyes are situated on top of a bizarre outgrowth on the carapace.

The common name for the entire family refers to the crab-like position of the legs, which allows the spiders to run sideways or even backward. The hind legs (III and IV) are shorter and gain footholds, whereas the longer and much stronger forelegs (I and II) are used to grab and retain prey. When a spider sits on a blossom and an insect

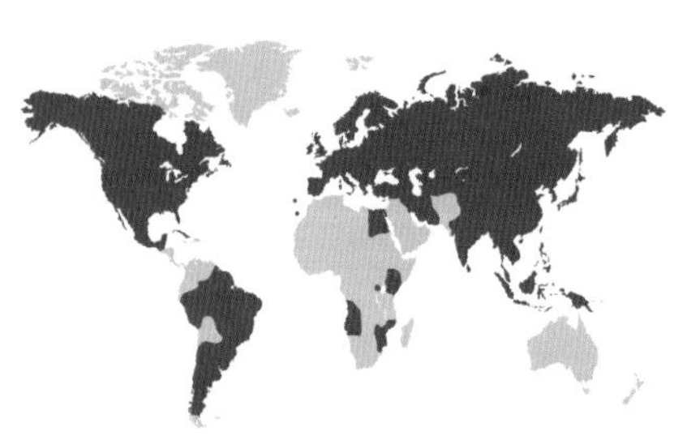

GENUS
Misumena

DISTRIBUTION
Palearctic, Americas, Africa, Southeast Asia to New Guinea

HABITAT
Vegetation in various habitats, mostly in or close to blossoms of flowers in open habitats like meadows

CHARACTERISTICS
- Laterigrade legs, with legs I and II much stronger than posterior legs
- Ability to change color
- Ambushes insects in blossoms

lands there, the spider strikes with its forelegs. It immediately bites the insect close to its head and inject its venom to immobilize the prey. Many other spiders will then start chewing and kneading the prey, so that the digestive fluid emerging from the mouth enters the insect's body. Crab spiders, in contrast, only pierce their prey and then regurgitate the digestive fluid onto it. After a moment, their predigested liquid prey is sucked in, and the process is repeated until the feeding process is finished. Thus, the exoskeleton of the prey remains largely intact, enabling ecologists to analyze the prey spectrum of the species in question. Other groups of spiders that use a similar sucking technique, rather than chewing, are Pholcidae and Theridiidae.

BELOW | *Misumena vatia* female, sitting in a typical ambush position with the two pairs of forelegs widely spread, waiting for flying insects. The white form exhibits characteristic red patches on the abdomen, which may also occur in the yellow form.

THOMISIDAE: *AMYCIAEA*

ANT-EATING CRAB SPIDERS

BELOW | The crab spider *Amyciaea lineatipes* (right) mimicking the weaver ant *Oecophylla smaragdina* (left). Note the dark patch on the spider's abdomen imitating the insect's eye, the dark line on the abdomen resembling the gap between its head and thorax, and the first pair of legs held like the insect's antennae.

Ants are usually the dominant insect group in tropical rainforests, but they are fierce and are therefore avoided by many spiders. Other spiders mimic ants in appearance and behavior; in some instances the mimicry may be aggressive, as the spiders attack the ants in question. In order to mimic the morphology of ants in general, or of particular ant species, some morphological characteristics of the spiders are modified. Crab spiders typically have laterigrade legs, for example, thereby resembling crabs in appearance and movement. *Amyciaea*, in contrast, look like an ant, and have therefore lost the typical leg position as well as the strong distinction between forelegs and hindlegs. In addition, they have evolved a size (around 0.2 inches/5 mm body length), coloration, and behavioral traits similar to those of the ants, and have developed structures that spiders do not usually possess. *Amyciaea* exhibit two dark patches on the abdomen, resembling the conspicuous compound eyes of *Oecophylla* weaver ants. The two front legs are also raised in the air, imitating the antennae of an ant. Although these structures are at opposite ends of the body, the camouflage is seemingly accepted by surrounding animals.

GENUS
Amyciaea

DISTRIBUTION
Tropics of Africa (Sierra Leone, Ivory Coast), Asia (India, China to Malaysia, Singapore, Sumatra, Vietnam), New Guinea, Australia

HABITAT
Tropical rainforests, on foliage and twigs of trees and shrubs

CHARACTERISTICS
- Live close to ants and prey on them
- Mimics one ant species (*Oecophylla smaragdina*)
- Patches on abdomen mimic the ant's eyes

Amyciaea is a small genus containing five described species. In Asia, these spiders live in the foliage and on twigs of trees and shrubs, where the weaver ants build their nests. To avoid accidental collisions, individuals produce a silken thread and hang there at a safe distance from the ant path. The advantage of the aggressive mimicry to *Amyciaea* is a virtually limitless food supply. Ants are bitten in the head region and are immediately paralyzed by the spider's venom.

Amyciaea males sometimes exhibit bridging behavior when searching for a mate. Standing on the tips of their leg tarsi and raising their abdomen in the air, the spiders release a thread of silk that gets caught by the wind and air currents. This behavior is the same as that of a spider starting to balloon. A ballooning spider will emit the silk thread until it provides sufficient lift to allow the animal to release its grip on the substrate and become airborne. In contrast, *Amyciaea* species wait instead until the thread catches on a neighboring plant part, then use the silken zip line to cross over to the other plant. In so doing, the spiders avoid running on the ground, where many predators are hunting.

BELOW | A male *Amyciaea lineatipes* releasing silken threads for bridging gaps on the jungle floor in order to cover more ground and avoid potential predators.

SALTICIDAE: *MARATUS*

PEACOCK SPIDERS

RIGHT | A male peacock spider from Australia, with the abdominal flaps extended in courtship position.

BELOW | A male peacock spider from Australia, with the abdominal flaps in their normal, resting position.

Like other jumping spiders, peacock spiders (members of the genus *Maratus*, which includes 85 species) have excellent vision. They also display some of the most elaborate courtship behavior of any spiders. Although the females are relatively drab in appearance, the males are dramatically colored. Most of the colors are produced by pigments deposited in scales (modified hairs) on the top and sides of the abdomen, but to produce bluish colors the scales are equipped with arrays of embedded nanostructures. These nanostructures can apparently amplify light; the spiders have different photoreceptors for red, blue, green, and ultraviolet.

Males typically have brightly colored, flap-like extensions that are normally wrapped around the sides of the abdomen. Once thought to help the animals glide through the air while jumping, they have now been shown to be actively used during courtship. Males have a long and flexible pedicel, which allows them to elevate their abdomen, displaying its fancy coloration; the abdominal flaps can also be greatly expanded, radically widening the appearance of the spider. Color

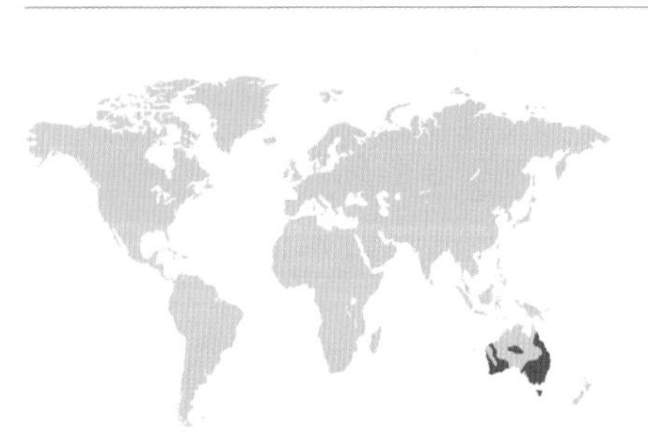

GENUS
Maratus

DISTRIBUTION
Australia

HABITAT
From semiarid areas such as sand dunes to temperate eucalypt forest litter

CHARACTERISTICS
- Males have abdominal flaps that are elevated during courtship
- Some males produce loud sounds during courtship
- Sharp color vision

spots, normally at the sides of the abdomen, then appear much closer to the center, often producing a face-like appearance. The males adopt a variety of poses, often elevating the third pair of legs into a V-shape, framing the abdomen, and crouching down on one side, deflecting the abdomen and extended legs in that direction.

In at least some species, males also signal to females by stridulation, rubbing the thickened bases of short, specialized setae found on the front of the abdomen across a pair of files situated on the back of the carapace. The movement produces sounds that can be heard by humans several feet away, and may be the loudest sounds produced by spiders.

SALTICIDAE: *PORTIA*

SPIDER-EATING JUMPING SPIDERS

BELOW | Frontal view of a female *Portia*, showing the tufts on the carapace and legs that help it resemble bits of detritus.

The genus *Portia* includes some of the most unusual jumping spiders. Unlike most other salticids, members of *Portia* can build their own webs, which are of two types. One is just a relatively small, horizontal platform on which the spiders rest (most salticids rest instead in silken retreats). The other is a much larger, three-dimensional structure, often built inside, or adjacent to, the webs of other spiders. The second web type is spun only by females. The spiders usually incorporate a dead leaf or piece of bark into these webs, which they maintain and use over several instars.

The bodies of *Portia* are highly ornamented, with tufts of hair on their legs, carapace, and abdomen. These, together with the spiders' mottled coloration, enable them to resemble pieces of detritus, especially in a resting position, where they fold their legs and palps close to the body. These spiders are able to capitalize on that appearance by moving very slowly and stealthily—so much so, that even other salticids they are slowly pursuing seem not to recognize them as an animal, much less a potential predator. They will sometimes even

GENUS
Portia

DISTRIBUTION
Africa, India to China, Australia

HABITAT
Rainforests

CHARACTERISTICS

- Unusual jumping spiders that build their own webs and invade those of other spiders
- Well-camouflaged body that resembles a piece of detritus
- Prefers to eat other spiders, but will also consume insects and spider eggs

ABOVE | Anterior view of a male *Portia* from Malaysia.

approach, or walk over or under, the *Portia*, as if it were a harmless clump of dirt.

The behavioral repertoire of *Portia* is extraordinarily diverse, however. Members of this genus are also adept at entering the webs of other spiders. Here, they slowly approach the owner, vibrating the web as they move, and sometimes cutting the threads as they go, restricting the mobility of their intended prey before making their final, quick lunge. Their venom seems to be unusually effective on other spiders. *Portia* will also act as kleptoparasites, attacking incompletely eaten insects caught in their hosts' webs as well as opening egg sacs and consuming their contents.

The protections of *Portia* against attacks from the spiders they hunt—and other predators—include a very tough cuticle, and legs and palps that are easily autotomized (detached by the *Portia* when attacked). This is clearly not an uncommon occurrence, as specimens found in nature are often missing some legs and/or palps.

SALTICIDAE: *MYRMARACHNE*

ANT-MIMICKING JUMPING SPIDERS

Jumping spiders constitute the largest family of spiders, with more than 6,000 species. It is thus not surprising that the family includes species that show substantial diversity in structure and behavior. Among the most obvious are several genera whose species include mimics of other kinds of animals. Some, such as *Coccorchestes*, found in New Guinea and Australia, closely resemble beetles. But perhaps the most striking are the more than 175 species placed in the genus *Myrmarachne*, all of which closely resemble ants. The adults of each species typically resemble one of the dominant species of ant found in their habitats; the resemblance is often striking, as shown in the photo below. Both the carapace and abdomen have constrictions that echo the body segments of the model species; even the hair patterns on the abdomen suggest the segmentation of an ant abdomen.

Nor is the resemblance confined to structure. The spiders move with their front pair of legs elevated, imitating the ants' antennae, and even bob their abdomens in much the same manner as ants do. The spiders also live in association

RIGHT | An ant-mimicking jumping spider of the genus *Myrmarachne*.

GENUS
Myrmarachne

DISTRIBUTION
Cosmopolitan; most speciose in tropical Africa, Southeast Asia, Australia

HABITAT
Grasslands

CHARACTERISTICS
- Closely resembles ants, both in morphology and behavior
- Carapace and abdomen have constrictions
- Males often have elongated chelicerae

ABOVE | A beetle-mimicking jumping spider of the genus *Coccorchestes* from New Guinea.

with ants of their model species, typically moving at the edges of foraging groups of ants. The resemblance does not confuse the ants; the spiders make no attempt to capture the ants as prey, avoiding direct contact (although they will opportunistically feed on ant eggs, larvae, and pupae). The mimicry in this case is thus not aggressive, but Batesian—meaning that the spiders gain protection from predators that have learned to avoid ants because of their noxious defenses.

Indeed, the life cycles of these spiders are closely connected with the ant faunas of their regions. Juveniles typically do not mimic the same ant species as the adults, but rather a different, smaller, and often very different-looking ant species that occurs in the same area. In at least one species, the ants will feed on the honeydew excreted by coccids (mealybugs), and the spiders will join the feast. Although spiders are often thought of as obligate predators (eating only live prey they capture themselves), some salticids will feed on flower nectar. One salticid species, *Bagheera kiplingi* from Mexico and Costa Rica, feeds primarily on Beltian bodies (specialized, food-rich leaf tips) produced by some plants. Another, *Evarcha culicivora* from Kenya, feeds indirectly on vertebrate blood, choosing as its prey only blood-fed female mosquitoes. Some tiny, web-building spiders may even obtain significant amounts of energy from plant pollen grains that adhere to the sticky silk in their webs and are ingested when the webs are taken down and the silk is consumed for recycling. These are exceptions, however: most salticids—indeed almost all spiders—are strict carnivores.

GLOSSARY

Terms in italics are cross-referenced;
s = singular; p = plural.

Abdomen: also called *opisthosoma*; the posterior part of a spider.

Apical: refers to the tip of a structure.

Apophysis (p = apophyses): a projection or appendage changing the general cylindrical or globular shape of a *sclerite*.

Book lungs: respiratory organs situated in front of (and also sometimes behind) the *epigastric furrow* on the ventral side of the *abdomen*, opening through narrow slits.

Bulbus (p = bulbi) or **bulb**: complex structure on distal segment of the *male palp*.

Calamistrum: a row of modified setae on *metatarsus* IV of *cribellate* spiders, used to comb the silk produced by the *cribellum*.

Carapace: the large *sclerite* covering the *cephalothorax*, the anterior part of the body.

Cephalothorax: also called the *prosoma*; anterior part of the spider covered by the *carapace*, bearing eyes, legs, and mouthparts.

Chelicerae (s = chelicera): first appendages of the cephalothorax, situated at the front of the carapace, consisting of a large basal *paturon* and a *fang*.

Claw tuft a dense group of hairs under the paired tarsal claws, typically well developed in hunting spiders; the tips of the hairs are expanded, bearing dozens of small projections that adhere to smooth surfaces.

Claws: see *tarsal claws*.

Clypeus: the space between the anterior edge of the carapace and the anterior eyes.

Colulus: short median protuberance in front of the spinnerets, considered a modification of the *cribellum*.

Coxa (p = coxae): see *legs* and *palp*.

Cribellate: provided with a *cribellum*.

Cribellum: spinning plate situated in front of the *spinnerets*, provided with many tiny, ribbed spigots; produced by the fusion of the anterior median spinnerets.

Cryptic species: a species that is well camouflaged to blend in with its background.

Cuspule: short, thick modified seta.

Distal: situated away from the center of the body.

Dorsum: upperside of the *abdomen*.

Ecribellate: without a *cribellum*.

Embolus (p = emboli): intromittent part of the male palpal *bulbus*, usually slender, sharp-tipped and strongly sclerotized, carrying the outlet of the sperm duct.

Endite: an expansion on the coxa of the palp, also called the maxilla or the gnathocoxa.

Entelegyne: describing spiders with an *epigynum*, provided with separate ducts for sperm transport during insemination (toward the spermathecae) and fertilization (toward the uterus); see also *haplogyne*.

Epigastric furrow: a large slit on the anterior part of the ventral side of the *abdomen*.

Epigynum (p = epigyna): a hard plate on the ventral side of the female abdomen on which the genital openings are located. It is only fully developed in adult females of *entelegyne* spiders; spiders belonging to Mygalomorphae and *haplogyne* Araneomorphae do not have an epigynum.

Eye formula: the position of the eyes is often expressed by digits separated by dashes (e. g., 2–2–2–2 means eyes in four rows of two eyes each; 6–2 means eyes in two rows, the first with six eyes, the second with two).

Fang: distal part of the *chelicera*, usually carrying the opening of the venom glands.

Femur (p = femora): see *legs* and *palp*.

Fovea: a central depression of the *cephalothorax*, often reduced to a dark longitudinal stripe; it corresponds to an internal ridge to which muscles are attached, also called the thoracic groove.

Gondwana: a supercontinent that existed between 550 million and 180 million years ago, and whose breakup resulted in the landmasses of South America, Africa, Arabia, India, Australia, and Antarctica.

Haplogyne: a term describing spiders that lack separate copulatory ducts; they thus have only one pair of ducts for transport of sperm from the uterus to the *spermathecae* during insemination, and back again to the uterus at fertilization; see also *entelegyne*.

Hemolymph: bloodlike fluid in the open circulatory system.

Holarctic: occurring in North America and Eurasia, thus restricted to the northern hemisphere.

Integument: external cuticular skin.

Labium (p = labia): sclerite situated between the *endites*, in front of the *sternum*.

Labrum (p = labra): upper lip; part of the mouthparts, hidden by the *chelicerae*.

Laterigrade: a term describing spiders that move sideways, like a crab, with the legs directed to the side (compare *prograde*).

Legs: consist of seven segments, the *coxa, trochanter, femur, patella, tibia, metatarsus*, and *tarsus*. Leg pairs are numbered I to IV from the front.

Male palp: refers to the modified tarsus of the palp in male spiders; it is the copulatory organ, although not in direct connection with the testes. It consists of an enlarged and hollow tarsus (cymbium), sometimes split in two (cymbium and paracymbium), which bears the copulatory *bulbus*; these vary enormously in shape and complexity, and are the most important characteristics for species identification in male spiders.

Metatarsus (p = metatarsi): see *legs*.

Opisthosoma: see *abdomen*.

Palearctic: occurring in temperate and subarctic regions of Europe, North Africa, and Asia.

Palp: also called *pedipalp*; second appendage of the cephalothorax in front of leg I, composed of *coxa*, *trochanter*, *femur*, *patella*, *tibia*, and *tarsus*. This appendage is modified into a copulatory organ in male spiders (see *male palp*).

Parthenogenetic: refers to species in which females can produce young without ever mating with a male.

Patella (p = patellae): see *legs* and *palp*.

Paturon: basal segment of the *chelicerae*.

Pedicel: narrow, waistlike connection between the *cephalothorax* and *abdomen*.

Pedipalp: see *palp*.

Petiolus: see *pedicel*.

Procurved: curved toward the front; for an eye row, this means that the lateral eyes are situated closer to the front than are the median eyes.

Prograde: describing spiders that have their legs directed forwards (I, II) and backwards (III and IV) (compare *laterigrade*).

Prolateral: refers to that side of an appendage that faces the front of the body.

Prosoma: the anterior part of a spider, including the *cephalothorax* and its appendages.

Recurved: curved toward the back; for an eye row, this means that the lateral eyes are situated closer to the back than are the median eyes.

Retrolateral: refers to that side of an appendage that faces the back of the body.

Sclerite: a single sclerotized part of the external hardened *integument*.

Scopula (p = scopulae): a brush of hairs on the ventral side of the terminal leg segments; improves grip on substrate or prey.

Scutum (p = scuta): sclerotized plate on the *abdomen*.

Serrula: a row of tiny teeth along the anterior margin of the endite.

Seta (p = setae): hairlike, tapered, and flexible structure on the legs and body (compare *spine* and *trichobothrium*).

Spermathecae (s = spermatheca): bladderlike, internal structures in female spiders that store sperm after insemination.

Spigots: tiny projections at the apex of the *spinnerets* through which the silk emerges.

Spine: pointed, articulated, rigid structure on the body and legs (compare *seta* and *trichobothrium*).

Spinnerets: silk-spinning appendages of the abdomen, usually arranged in three pairs: anterior lateral, posterior median, and posterior lateral. All are provided with small *spigots* through which the silk emerges.

Sternum: large sclerite on the ventral side of the *cephalothorax*, situated between the leg *coxae*.

Stridulating organ: a series of thin ridges on a sclerite, often on the lateral side of the *chelicerae*, forming a stridulating file; it corresponds with a series of short, stiff setae or cusps on another part of the body that can run along the file to make a stridulatory sound.

Synanthropic: living in or around human habitations, and thus easily transported by humans to new areas.

Tapetum (p = tapeta): light-reflecting crystalline structure behind the retinal cells of the secondary eyes.

Tarsal claws: situated at the tip of the tarsus. There is either a single pair, often hidden in a claw tuft, or a pair and a third single claw, which is much smaller and situated under the paired claws; these claws are usually pectinate (provided with a row of teeth).

Tarsal organ: a receptor, most often a tiny depression on the dorsal side of the tarsus. In some families this organ is well developed, sticking out above the surrounding *integument* and easily visible.

Tarsus (p = tarsi): see *legs*.

Tibia (p = tibiae): see *legs* and *palp*.

Tibial apophysis: an *apophysis* on the tibia of the male palp.

Tracheae (s = trachea): thin, sclerotized internal tubes, part of the respiratory system in many Araneomorphae. They open onto the *venter* of the *abdomen* through one or more *tracheal spiracles*.

Tracheal spiracle: a small slit on the ventral side of the *abdomen*, usually immediately in front of the *spinnerets*.

Trichobothrium (p = trichobothria): supple, hairlike structure on legs and palps implanted in a shallow alveolus (compare *seta* and *spine*), sensitive to air movement.

Trochanter: see *legs* and *palp*.

Venter: bottom side of the *abdomen*.

FURTHER READING

SPIDER BIOLOGY & NATURAL HISTORY

Barth, Friedrich G. 2002.
A spider's world: senses and behavior. Springer.

Brunetta, Leslie and Catherine L. Craig. 2010.
Spider silk: evolution and 400 million years of spinning, waiting, snagging and mating. Yale University Press.

Craig, Catherine L. 2003.
Spiderwebs and silk: tracing evolution from molecules to genes to phenotypes. Oxford University Press.

Foelix, Rainer F. 2011.
Biology of spiders. Third edition. Oxford University Press.

Herberstein, Marie (ed.). 2010.
Spider behaviour: flexibility and versatility. Cambridge University Press.

Jocqué, Rudy, and Ansie S. Dippenaar-Schoeman. 2006.
Spider families of the world. Royal Museum for Central Africa.

Kelly, Lynn. 2009.
Spiders: learning to love them. Allen & Unwin.

Nentwig, Wolfgang (ed.). 2013.
Spider ecophysiology. Springer.

Penney, David (ed.). 2013.
Spider research in the 21st century. Siri Scientific Press.

Shear, William A. (ed.). 1986.
Spiders: webs, behavior, and evolution. Stanford University Press.

Wise, David H. 1993.
Spiders in ecological webs. Cambridge University Press.

TAXONOMIC LITERATURE

The taxonomic literature on spiders is indexed in the World Spider Catalog (at https://wsc.nmbe.ch). Anyone can join the World Spider Catalog Association, at no expense, which enables open access to pdf versions of the relevant publications.

REGIONAL & FIELD GUIDES

NORTH AMERICA

Bradley, Richard A. 2012.
Common spiders of North America. University of California Press.

Ubick, Darrell et al. (eds). 2017.
Spiders of North America – an identification manual, second edition. American Arachnological Society.

EUROPE

Bee, Lawrence, Geoff Oxford, and Helen Smith. 2017.
Britain's spiders: a field guide. Princeton University Press.

Bellmann, Heiko. 2006.
Kosmos Atlas Spinnentiere Europas [in German]. Kosmos Verlags.

Roberts, Michael J. 1998.
Spiders of Britain and northern Europe. Harper Collins.

ASIA

Koh, Joseph K. H. and Nicky Bay. 2019.
Borneo Spiders: A Photographic Field Guide.
Sabah Forestry Department.

Ono, Hirotsugo and Kiyoto Ogata. 2018.
Spiders of Japan: their natural history and diversity [in Japanese].
Tokai University Press.

Zhang, Zhisheng and Luyu Wang. 2017.
Chinese spiders illustrated [in Chinese]. Chonqing University Press.

SOUTH AMERICA

Aisenberg Anita, Carlos A. Toscano-Gadea, and Soledad Ghione. 2011.
Guía de Arácnidos del Uruguay [in Spanish]. Ediciones de la Fuga.

AFRICA

Dippenaar-Schoeman, Ansie S. 2014.
Field guide to the spiders of South Africa. Lapa Uitgewers.

Lerot, Astri and John. 2003.
Spiders of southern Africa. Random House Struik.

AUSTRALIA

Framenau, Wolker W., Barbara C. Baehr, and Paul Zeborowsky. 2014.
A guide to the spiders of Australia. CSIRO Publishing.

Whyte, Robert and Greg Anderson. 2017.
A field guide to spiders of Australia. CSIRO Publishing.

ABOUT THE AUTHORS

NORMAN I. PLATNICK

Norman I. Platnick is the Peter J. Solomon Family Curator Emeritus of Spiders at the American Museum of Natural History, where he curated the world's largest collection of spiders. A Ph.D. recipient from Harvard University, Platnick has described more than 1,800 species of spider, making him the second most prolific arachnologist in history. He originated the World Spider Catalog, now an Internet database that provides workers around the globe with ready access to all the taxonomic literature on more than 48,000 species, a resource that is unmatched for any other group of organisms. He led the Planetary Biodiversity Inventory project on goblin spiders, an international collaboration of 45 investigators that jointly tripled our knowledge of the Oonopidae, and is currently working on guardstone spiders (Phrurolithidae).

GUSTAVO HORMIGA

Gustavo Hormiga is the Ruth Weintraub Professor of Biology at the Department of Biological Sciences of the George Washington University, in Washington, D.C. He completed his undergraduate studies at the Universitat de Barcelona and his Ph.D. at the University of Maryland. His research focuses on the systematics and evolution of spiders and other arachnids, with an emphasis on orb-weavers and their relatives. He has trained numerous doctoral students from around the world in spider systematics. Hormiga is also a Research Associate of the Museum of Comparative Zoology at Harvard University, the American Museum of Natural History in New York, and the California Academy of Sciences (San Francisco), where he is also an Elected Fellow, and a Visiting Professor at the University of Copenhagen. He has done extensive fieldwork around the world to study spiders, and he has discovered and described numerous new species.

PETER JÄGER

Peter Jäger is Scientific Curator of Arachnida and Myriapoda at the Senckenberg Research Institute and Nature Museum in Frankfurt, Germany, where the world's fifth-largest natural history collection is located. Jäger is a specialist on the spider family Sparassidae (huntsman spiders) and has also worked on various other spider taxa. In his 230-plus publications he has described more than 400 new spider species. Jäger is a member of the editorial board of the World Spider Catalog, and of the journals *Arachnologische Mitteilungen* and *Acta Arachnologica Sinica*, and is an associate editor of *Zootaxa*, a journal describing about a quarter of all animal species. In 2012 Jäger founded the Asian Society of Arachnology and is still involved in many collaborations in Asia. His latest project is a revision of *Olios*, the largest genus of the huntsman spiders.

RUDY JOCQUÉ

Rudy Jocqué is honorary head of the invertebrate non-insects section at the Royal Museum for Central Africa in Tervuren, Belgium. He was senior scientist at the University of Ghent and produced a thesis on the spider fauna of heathland in Belgium. He spent three years in tropical Africa as an expert for the Food and Agriculture Organization, where he began his studies on Afrotropical spiders. He carried out fieldwork in some 15 African countries. His main contributions to arachnology are a revision on a global scale of the Zodariidae, and textbooks on African spiders and on spider families of the world in collaboration with A. Dippenaar-Schoeman. He has described more than 800 new spider taxa, mainly in the Zodariidae, and contributed to the taxonomy of more than 10 families. He discovered and named an external sclerite of spiders—the chilum—that is now considered an essential part of spider descriptions.

MARTÍN J. RAMÍREZ

Martín J. Ramírez is a research scientist at the National Scientific and Technical Research Council of Argentina, and Curator of Arachnids at the Museo Argentino de Ciencias Naturales Bernardino Rivadavia, where he curates one of the largest spider collections in Latin America. He graduated from Buenos Aires University, where he currently teaches as an adjunct professor. Ramírez has described more than 200 species of spider and specializes in the morphology and phylogeny of araneomorphs. He was the lead organizer of the Spider Anatomy Ontology, a structured online glossary of anatomical terms aimed at interdisciplinary research in spiders. He is currently working on ghost spiders, the higher-level phylogeny of spiders using molecular and morphological data, and tracing the evolution of organ systems in spiders.

ROBERT J. RAVEN

Robert J. Raven is Principal Curator of Arachnida at the Queensland Museum, Australia. He has described over 450 new species, including more mygalomorphs than any person and he documented over 300 mygalomorph genera during his post-doctoral year at the American Museum of Natural History. He has trained students from Brazil, Israel, Mexico, India, Thailand, Germany, and the USA as well as Australia. He converted the then text-based World Spider Catalog to a relational database, which he managed until 2010. In 1992, he hosted the first International Congress of Arachnology in the southern hemisphere. His present focus, as well as Australian tarantulas and other mygalomorphs, includes the Miturgidae and related ground-hunting spiders, a rich group that has diversified extraordinarily in Australia's outback.

INDEX

PICTURE CREDITS

The publisher would like to thank the following for permission to reproduce copyright material:

Adam Parsons: 37, 39, 41, 46, 216. **Alamy Stock Photo** Andrew Mackay: 76; Avalon/Photoshot License: 132, 153 (inset), 202; blickwinkel: 4, 7TL (D), 16, 21TR, 35, 69, 112, 113, 131, 194, 217; ephotocorp: 210; FLPA: 33; Gillian Pullinger: 225BL; Hemis: 122; João Burini: 197; Juan Antonio Guerrero Escobar: 42; Keith Davis: 65; Kjell Sandved: 6BL; Minden Pictures: 48, 49, 121, 201, 249; Nick Upton: 145; Nic van Oudtshoorn: 117; Papilio: 86; Peter Yeeles: 247; Premaphotos: 183B; RGB Ventures SuperStock: 116; Stefan Sollfors: 142, 230. **Alex Hyde**: 87. **Alice Abela**: 79. **Alvaro Laborda**: 80. **Annika Lindqvist**: 36. **Arnaud Henrard**: 138, 170, 174, 229.**Arthur Anker/Pedro H. Martin**: 220. **Bryce McQuillan**: 13B, 43, 70, 193, 213. **Cédric Alonso**: 72. **Cedric Lee**: 93T. **Danny Declercq**: 206, 240 **Dreamstime/Tamara Lee Harding**: 199. **Eric Carpenter**: 191. **Felix Fleck @felixfleckonline.com**: 28. **Frank Hecker Naturfoto**: 188; /Heiko Bellmann: 162, 168, 169, 179, 203. **Gabriel Pompozzi**: 26, 44. **Getty Images/Auscape/ Universal Images Group**: 156. **Gonzalo Giribet**: 99. **Greg Anderson**: 139. **Gustavo Hormiga**: 62, 71, 100, 106, 107, 120, 123, 124, 127, 133, 134B, 135L, 140, 141, 144, 147T, 149, 157, 159, 161. **Hans Henderickx**: 151, 224. **Ian C. Riddell**: 171. **Ingo Rechenberg**: 236, 237B. **James Koh**: 74. **Jan Bosselaers**: 160, 207. **Javier Replinger (jotafoto)**: 184. **Jeremy Miller**: 111. **João P. Burini**: 148. **Joel Ledford**: 73, 85. **John Leroy**: 218. **Joseph K. H. Koh**: 126. **Jürgen Otto**: 244, 245. **Kamil Stajniak**: 143. **Karl Granzien**: 125. **Kristi Ellingsen**: 61. **Lara Lopardo**: 134T. **Leonardo Sousa Carvalho**: 200, 219. **Lex Peeters**: 110. **Ludwig Jansen (https://ludwig .piwigo.com)**: 105, 167, 225BR, 226. **Marshal Hedin**: 23, 24, 27, 31, 58, 64, 78, 81, 158, 175, 190. **Martin J. Ramirez**: 10–11 (A–H), 12 (I & J), 13TL (K), 13TR (L), 15TR, 15CTR, 15B, 57, 60, 63, 84, 101, 222, 223, 231. **Matias Izquierdo**: 92BR (courtesy of AMNH), 195. **Merlijn Jocque**: 208. **Michael Pankratz**: 129. **Nathan George Burgess Russ Proper Job Productions**: 227. **Nicky Bay**: 51, 66, 68, 83, 96, 114, 115, 118, 119, 130, 152, 153 (main image), 172, 173, 196, 205, 233, 235, 242, 248. **Norman I. Platnick/courtesy of AMNH**: 91, 93B, 95. **Oscar Mendez**: 214. **Peter Jäger**: 7BR, 67, 75, 82, 204, 212, 232, 234, 237T, 243. **Peter Webb**: 228. **Pierre Anquet**: 185. **Pierre Oger**: 146, 147B, 150. **Queensland Museum**/Bruce Cowell 45, 50; G. May: 186; Robert J. Raven: 21TL, 29, 30, 47, 177, 180, 181, 187. **Robert Whyte**: 34, 40, 92TL, 94, 98, 192, 215T. **Rogerio Bertani**: 32. **Ron Atkinson**: 215 (inset). **Ryan Kaldari**: 59. **Shannon Christensen**: 25. **Shuqiang Li**: 135R. **Shutterstock/**3DMI: 7C-UP-LE (F); Alex Stemmer: 53; Anna Seropiani: 189; Cathy Keifer: 8B; Cornel Constantin: 137; Dennis van de Water: 198; Dev_Maryna: 164; Ernest Cooper: 6CR (C), 54, 89; Ezume Images: 6TR (B); D. Kucharski K. Kucharska: 136; Gallinago_media: 7TR (E); Gerry Bishop: 22; IanJHall: 183T; IanRedding: 182; James van den Broek: 38; JorgeOrtiz_1976: 128; Karel Gallas: 109; Keith Davis: 65; Korovin Aleksandr: 239; Melinda Fawver: 15 CLR; Mihai_Andritoiu: 88; Milan Zygmunt: 18; OoddySmile Studio: 7CBL (G); PetlinDmitry 6CL (A); Pong Wira: 17; RealityImages: 211; Sebastian Janicki: 5; serg_bimbirekov: 7CR (H); Space creator: 77; Svitlyk: 241; thatmacroguy: 238; XH2: 3, 15TL, 52; Wanchat M: 246. **Siel Wellens**: 209. **Steve O'Shea**: 176. **Stuart Schaum**: 221. **Thomas J. Astle**: 97. **W. Uys**: 108. **Walter P. Pfliegler**: 90, 163.**Wikimedia Commons/**Didier Descouens (CC BY-SA 3.0) (https://commons. wikimedia.org/wiki/File:Araneus_diadematus_MHNT_Femelle_ Fronton.jpg): 155; Flickr/Alejandro Santillana: 102; Sarefo (CC BY-SA 4.0) (https://commons.wikimedia.org/wiki/File:Tegenaria. atrica.creating.eggsac.1.jpg): 178.

Illustrations by **John Woodcock** (© The Ivy Press): 8T, 9, 20, 56, 104, 166.

All reasonable efforts have been made to trace copyright holders and to obtain their permission for the use of copyright material. The publisher apologizes for any errors or omissions and will gratefully incorporate any corrections in future reprints if notified.